Celebrating God's Cosmic
Perichoresis

Celebrating God's Cosmic *Perichoresis*

The Eschatological Panentheism of Jürgen Moltmann as a Resource for an Ecological Christian Worship

By Bryan Jeongguk Lee

Foreword by Charles Fensham

PICKWICK *Publications* · Eugene, Oregon

CELEBRATING GOD'S COSMIC PERICHORESIS
The Eschatological Panentheism of Jürgen Moltmann as a Resource for an Ecological Christian Worship

Copyright © 2011 Bryan Jeongguk Lee. All rights reserved. Except for brief quotations in critical publications or reviews, no part of this book may be reproduced in any manner without prior written permission from the publisher. Write: Permissions, Wipf and Stock Publishers, 199 W. 8th Ave., Suite 3, Eugene, OR 97401.

Pickwick Publications
An Imprint of Wipf and Stock Publishers
199 W. 8th Ave., Suite 3
Eugene, OR 97401

www.wipfandstock.com

isbn 13: 978-1-60899-908-8

Cataloging-in-Publication data:

Lee, Bryan Jeongguk.

    Celebrating God's cosmic perichoresis: the eschatological panentheism of Jürgen Moltmann as a resource for an ecological Christian worship / Bryan Jeongguk Lee ; foreword by Charles Fensham.

    p. 196 ; 23 cm. — Includes bibliographical references.

    ISBN 13: 978-1-60899-908-8

    1. Moltman, Jürgen—Contributions in Christian doctrine of human ecology. 2. Human ecology—Religious aspects—Christianity—History of doctrines—20th century. I. Fensham, Charles James. II. Title.

BT695.5 .L43 2011

Manufactured in the U.S.A.

To my beloved wife,
Christine Jeonghee Ryu,
who has shared all the difficulties and joys of my journey

Contents

Foreword / ix
Introduction / xiii

1. Reformed Standards for Envisioning a Viable Christian Ecological Theology / 1
2. Trinitarian History and Its Glorification in the Cosmos / 25
3. The Cross and Resurrection as the Foundation for Ecological Redemption / 67
4. Christian Worship as Anticipatory Celebration and Missional Sending / 95
5. Envisioning an Ecological Reformed Worship / 137

   Conclusion / 151

   Bibliography / 157

Foreword

THERE IS NO DOUBT that climate change is a reality and that the earth's biosphere is in grave danger. As I am writing this forward it is reported on the news that a thorough scientific review of the data and research findings of the scientists at East Anglia University in the United Kingdom has been shown to have integrity. This is despite the uncertainty cast over their work in the so called "climate gate affair" over the last year due to leaked email conversations that suggested that they were less than open to listen to their critics. Some of the more pessimistic prognosticators, such as James Lovelock, believe that the human contribution to these problems have already reached the point of no return. Africa, already the cradle of the "bottom billion" of humankind in terms of poverty, is the continent affected most with arable land dramatically shrinking away and creating the potential for large numbers of hungry refugees looking for food. North America and Europe and to a growing degree Asia are contributing most to the gasses spewing into our frail global atmosphere that are exacerbating this problem. Often Christian congregations of all stripes in Europe and North America pay scant attention to their own entanglement with pollution. Over the last three decades several credible Christian theological responses to the ecological crisis have been developed. These include the creative work of Thomas Berry and the challenging perspectives of eco-feminist theologians. Nevertheless, even though these perspectives have gained some traction in European and North American communities they have not yet turned into a larger movement committed to bring faith and ecology together.

Often the blame for ecological destruction is rightly placed at the feet of the Christian church in the West and the reigning theology of human confidence in our ability to take control of creation. Moreover the blame is taken right back to the Scriptures that play such an important role in the Judeo-Christian tradition. Does the first chapter of Genesis not direct humankind to have *dominion* over the earth and to *subdue it*? To make things worse, texts such as Genesis 1 use the strongest possible exploitative language. It has become the common practice of those advocating for environmental awareness to critique these traditions as they reached their zenith in Renee Descartes' famous claim that we are *"maitres et possesseur de la creation."* Often such critiques lead to the wholesale rejection of the larger Christian tradition and the story of creation as told in the Book of Genesis. As a Christian one is thus presented with the alternative of reading against the biblical text, or selectively rejecting certain texts. Such critical approaches, as creative and sincere as they might be, do not receive much traction in Christian communities that take the study of the biblical text seriously. Is there then a way of reading the Bible for ecology? Is there a way through the text to a theological vision that is inspiring for Christians of those traditions that take Bible study and wrestling with the text in all its complexity and cultural structures seriously?

In this book, Dr. Bryan Jeongguk Lee takes this challenge seriously. He argues that the church itself has to come to terms with its less than laudable history in caring for the earth. But he believes that the theology of Jürgen Moltmann provides exactly the kind of theological vision that can function in Christian communities that take both the Bible and its interpretation seriously, through rather than against the text. In the process, Dr. Lee makes accessible the wonderful inspiring theological vision of Jürgen Moltmann for understanding God's act of creation. He shows how Moltmann presents us with a way to understand the place of the earth and humankind in creation within the greater unfolding eschatological understanding of God's purpose with creation. God is both Creator apart from creation and also intimately involved with it, thus, this perspective is described as *panentheistic* as distinguished from pantheism—God and creation is one, or pure theism—God is always completely separate from creation. Moltmann's vision responds to the great questions of theodicy, the holocaust, and now the ecological disaster facing us. Not only does Dr. Lee accurately depict Moltmann's inspiring ideas, he also makes them immanently accessible to the reader by way of metaphor and exploration. At the same time he adds his own layers that build on Motlmann's fruitful unpacking of the larger Christian tradition.

Rather than discarding tradition, as some would want us to do, Dr. Lee together with Moltmann critically reads through and with the text and tradition, particularly Western Protestant tradition, and shows how the enlightenment hijacked and warped a wholesome understanding of human stewardship under God. Instead of inspiring humankind to be caring human gardeners, the Enlightenment and in the Industrial Revolution turned Christian understanding of the Biblical text into a self-focused quest for mastery and control over creation. The book is worth reading simply for this enlightening perspective. Moreover, as someone from the Asian Presbyterian-Reformed context, Dr. Lee dialogues with that tradition as he works towards building an ecological theology of worship that can inspire and empower churches to engage actively in ecological responsibility. In the process he engages the sacramental theologies of Luther and Calvin and offers both a critique and constructive proposal.

For Dr. Lee it is not enough to make a good and credible theological argument; the project will not be complete until local Christian churches absorb a vision for the healing of creation in responsibility before God in the weekly and daily practice of worship. After all, doxology is one of the great constants of the church through the ages! We sing what we believe and we come to believe what we sing. Our doxology as communities of faith lies at the heart of our practice. It is one thing then to become theologically convinced of our responsibility before God for the earth and even outer space that is increasingly being polluted by human junk, it is another thing when local Christian communities sing, pray and speak with conviction about this responsibility as they move to act. Dr. Lee believes our worship must be touched by this profound theological vision and awareness of responsibility, and he offers both a convincing and inspiring-poetic vision and a practical perspective that challenges us to examine our prayers, our hymns, and our daily Christian practice in the light of Moltmann's ecological panentheism.

I love this approach! Too often theology stays only in the cerebral sphere or on the page of a book. How we worship and its content are not incidental or mere cultural convention, even though culture is an important part of it. How we worship and the content of our worship matter because they shape the minds with which we are called to love God in the great commandment. When we can move, beyond romanticised hymns for creation that merely paint its beauty, to a place where we can confess our culpability and take responsibility for creation, then we will be able to turn around and recapture the true meaning of our core task—the core covenant of God—that calls us to tend and care for the garden. Of course we need to continue to sing the beauty of creation, but, today, we also need a great deal of lament.

In as much as we see ourselves as righteous exploiters of creation—masters and possessors—no matter how we may dress it up religiously, we stray from our accountability to God. Since "the earth is the Lord's," we are responsible to God for what we do with it. At present we are killing God's earth. Yet, the great covenant of creation calls us to care, tend, name and look after creation. We are to be gardeners, and garden dwellers. There is no more important task given to humankind. Given the broken state of this earth that reverberates through creation we are called to turn back to the task of world repair. We are called to turn to God's great redemptive plan for creation. We learn in Colossians 1, when the author quotes an early Christian hymn that those early Christians did not sing only of their own reconciliation or that of the souls of the lost, but that they sang of the reconciliation of all things in heaven and earth with God. This book calls us to sing this song anew as a repentant people who have strayed far from God's purpose. This book is right. Please read it with great care and then write a hymn of creation in action and word.

Charles Fensham
July, 2010.

Introduction

THE RESEARCH FIELD

THE PURPOSE OF THIS thesis is two-fold. First, this thesis aims to present Jürgen Moltmann's eschatological panentheism as a viable ecological theology for today. Second, this thesis aims to link Moltmann's ecological theology to Christian worship, particularly but not limited to Reformed worship, in order to foster Christian ecological stewardship.

Therefore, the research field of this thesis would be the ecological aspects of Moltmann's theology as well as his theology of worship. The thesis will also explore the liturgical possibilities of Moltmann's eschatological panentheism for an ecological Christian worship.

BACKGROUND OF THE THESIS

In the face of today's unprecedented ecological crisis, Christianity has been under severe attack from those who are concerned with ecology and those who want to imagine a new ecological way of life. For them, Christianity not only shares in the guilt of causing today's environmental crisis, but also is unwilling and incapable of providing any help in re-envisioning the required new way of life on earth.

H. Paul Santmire's historical study of the Christian tradition reveals its ambiguous ecological legacy. Although a considerable amount of the earlier tradition, as well as the advent of the ecological legacy of Saint Francis of Assisi, might be welcome music to ecological ears, Santmire lays bare how modern Christian theology demoted the natural world from its natural horizontal position to a mere profane and dull backdrop against which a vertical theanthropological (God-human) drama of salvation unfolds.[1]

Especially, Christian apocalyptic eschatology, with its assumption of a separate origin and destiny of humankind from the rest of the world and the entailing world-deserting eschatological vision, forms a spirituality that is fundamentally insensitive and indifferent to nature.

In this regard, Christianity seems hopeless in providing an ecological vision, an ecological spirituality, and an ecological way of life in which a more inclusive and sensitive relationship is established between humanity and the natural world, where humans exist as a part but not the end-all of the natural world. As a result, ecologists tend to look to Eastern and Native American religions for the kind of ecological vision required for today's world.[2]

Christian theology, however, has not stood totally aloof. In fact, Christian theologians have responded with various strategies and degrees of success, emphasising the ecological aspects of the Christian tradition and establishing an ecological theology.

Harold Wells identifies the common goal of a wide spectrum of ecological theologies as providing a theological antidote to the profanation of the natural world.[3] Wells maintains that, despite a great diversity, from new cosmologists to Evangelicals for Social Action, today's ecological theologians share two common themes or stances: it is necessary 1) to recognise the divine immanence in the creation, and 2) to reposition humans as one of many mutually-dependant creatures in the ecological web.[4]

---

1. Santmire, *Travail of Nature*.

2. Thomas Berry is one of the first ecologists who have appreciated religious resources for ecological way of living outside Christianity. See, among others, his "Catholic Church and the Religions of the World," in *Riverdale Papers X*, 8–9. Regarding the contributions, contradictions, and limitations of Asian religions including Hinduism, Buddhism, Confucianism, and Taoism, see Jai-Don Lee, "Towards an Asian Ecotheology," 96–131. Mary Evelyn Tucker also wrote or edited extensively on Asian and native American religions' ecological potential and their challenge to modern concepts of nature and culture; among many of her writings, see Chapple and Evelyn Tucker, *Hinduism and Ecology*; Tucker and Berthrong, *Confucianism and Ecology*; Tucker and Williams, *Buddhism and Ecology*. On Native American religions, see Grim, "Living in a Universe," 243–60; Grim, *Indigenous Traditions and Ecology*.

3. Wells, "Flesh of God," 56–57.

4. Ibid.

Panentheism, then, becomes important for ecological theology because of its ability to affirm God's universal indwelling in creation. This overcomes the anti-ecological, one-sided emphasis of modern Christian theology on God's transcendence over and against the material universe, while at the same time avoids a pantheistic identification of God with the universe, thus circumventing idolatry.

Employing panentheism, however, is not without problems. As Wells points out in his criticism of Sallie McFague's panentheistic vision of the world as "Body of God," panentheism runs the risk of divinising nature and, by extension, ratifying whatever way of living that already is established in human society. Moreover, it imposes a very important christological question: how can we differentiate the special presence of God in Jesus and the universal presence of God in the cosmos?[5]

How, then, can Christian theology provide today's world with a viable Christian ecological vision that does not de-center Jesus Christ nor divinise nature?[6] Moreover, how can we go beyond just talk? Does theology really matter in humanity's contemporary race against time? How can Christian theology effectively galvanise the public and equip them with an ethical energy to waylay our current path of self-destruction?

I would argue that Jürgen Moltmann's "eschatological panentheism," along with his theology of worship, can be a very significant Christian contribution to these questions. In this thesis, I will demonstrate the ecological potential of his eschatological panentheism in relation to his concept of Christian worship as an anticipatory celebration of the eschatological panentheistic Sabbath.

Because of his eschatological-theological articulation of Trinity, Christology, pneumatology, and ecology, Moltmann is one of the central figures in today's Christian theology. He has also greatly heightened the theological importance of ecological awareness with his emphasis on, and challenge of, eschatology. Moltmann's eschatological panenthe-

---

5. Wells criticizes McFague that her proposal on the radicalization of the incarnation doctrine is in fact domestication of the doctrine, deserting the centre of Christian theology. Ibid., 61.

6. Divinization of nature tends to legitimize whatever already is; everything should be regarded a legitimate expression of divine will, and thus, preserved as sacred. This stance resists any kind of change in the individual and collective human way of life, representing political conservatism. A viable Christian ecological vision should be able to present a persuasive case for a fundamental change not only of our perception of our relationship toward nature but of the ways we live that impact nature.

ism—God's being all in all at the end of time—is a perichoretic union of God and creation. Here, the concept of *perichoresis*, a key term for Moltmann in defining the relationship among the three divine Persons, is applied to describe the eschatological relationship between God and the whole cosmos.

Moltmann's theology, however, is not coloured by day-dreaming of a distant future. Rather, his theology is among the strongest forms of post-holocaust theologies, thoroughly grounded in, and acknowledging, the grim and bitter reality of today's unredeemed world. His messianic Christology and his concept of Trinitarian dealings with the world stand between today's misery and struggles toward liberation and the eschatological bliss of eternal union with God.

In fact, the whole of Moltmann's theology is placed in this tension and geared toward the dissolution of this tension between the promised future and present reality of "not yet." Moltmann's processive Christology, for example, is a story of the man who came as a messianic promise, who died in union with the outcasts of the world and everything transitory and mortal, but who was resurrected as an eschatological promise. He is not yet a *pantocrator*; He is still on His way to His Sonship which will be consummated when He hands over the Reign to God the Father (1 Cor 15:24–28), so that God may be all in all to form a perichoretic interpenetration called God's eschatological Sabbath.

Moltmann places worship in this unresolved tension of hope and agony, of "already" and "not yet". Worship is God's workshop where worshippers are transformed into the image of God whom they worship. In worship, through and by the presence of the One who is to come and anticipated taste of ecological bliss of Eschaton, a moral will to fight against the present destructive current will be formed and moral imagination on how to live out the eschatologically-perceived reality will be ignited. In union with the crucified One and in the inviting and liberating power of the Holy Spirit, Christians enter into the Trinitarian bond of love and appropriate the eschatological event of Christ's resurrection into worship. In this context, worship goes beyond remembering Christ's past and anticipates the eschatological future feast. In this structure of remembrance and hope that becomes possible by the power of the Holy Spirit, Moltmann sees a great mission epicenter for the coming Reign of God. This is why I believe Moltmann's eschatological panentheism—alongside his concept of worship as anticipatory feast of the eschatological panentheistic Sabbath—could play a crucial role in nurturing an ecological spirituality in the church and equipping it with ethical direction and missionary energy.

Moltmann himself, however, does not go further to link his understanding of Christian worship mainly developed in *The Church in the Power of the Spirit* with his ecological concern mainly expressed in his later books such as *God in Creation, The Way of Jesus Christ, The Spirit of Life*, and *The Coming of God*. His messianic ecclesiology defined in terms of anticipatory participation of the church in God's eschatological consummation of history is not reviewed and re-coloured in terms of his later ecological concern and his all-embracing eschatological concept of New Heaven and New Earth. Accordingly, the liturgical implications of such an ecological eschatology have yet to be pronounced. One of the purposes of this thesis is seeking to open a constructive dialogue between theology and liturgy toward an ecological worship by offering, based on Moltmann's eschatological panentheism, some theological principles and suggestions toward a future ecological liturgy.

THESIS STATEMENT

Jürgen Moltmann's eschatological panentheism is a viable Christian ecological theology for today and offers significant theological resources for awakening ecological awareness in Christian worship.

PARAMETERS AND ASSUMPTIONS

When it comes to today's Christian ecological theology, it is beyond the scope of this thesis to explore and evaluate all approaches. Here I identify four distinctive approaches in contemporary ecological theology: Thomas Berry's cosmological approach; John Cobb's metaphysical approach based on Whiteheadian process thought; Rosemary Ruether's eco-feminist approach; and Leonardo Boff's eco-liberationist approach.[7]

Although I appreciate all of these approaches, especially the contribution Thomas Berry made in his presentation of cosmogenetic understanding of the world in dialogue with the contemporary scientific

---

7. Jeong-Woo Lee, "Toward a Trinitarian Ecological Theology."

cosmology, I will take a Reformed approach to this subject of ecological theology and use its standards.

I will try to articulate, in the terms of Christological and Trinitarian Gospel, the ecological hope found in Christ, especially in the resurrection of the crucified One and in the ensuing eschatological outlook as a logical cosmological extention of that resurrection. The Scriptures tell us that Jesus Christ "has become for us wisdom from God—that is, our righteousness, holiness and redemption" (1 Cor 1:30). The Barmen Declaration also emphatically states that **"Jesus Christ, as He is attested for us in Holy Scripture, is the one Word of God which we have to hear and which we have to trust and obey in life and in death."**[8] This stance reaffirms the Reformed tradition's caution regarding the sin of idolatry and its emphasis on the lordship of Jesus Christ. In this regard Harold Wells appropriately points out: "If we decide theologically that we should be centered somewhere else than in Jesus Christ, if we conclude that our norm for Christian theology is to be found elsewhere, then indeed our language of worship and Christian life must change drastically."[9]

In line with this Christ-centerd thinking, I also believe that it is nowhere else than in Jesus Christ that Christians can and should find an ecological hope and missionary energy for ecological transformation. Accordingly, my approach to envisioning a viable Christian ecological theology and ecological Christian worship will be Christ-centerd and Trinitarian.[10] This is also Moltmann's decisive approach. For him, "Christology is only the beginning of eschatology; and eschatology, as the Christian faith understands it, is always the consummation of Christology."[11] Therefore, by following a christological-Trinitarian way of thinking, it becomes possible to decisively link our faith in Jesus Christ with the ecological hope expressed in the eschatological terms of the New Heaven and the New Earth.

Of eschatology's many strains, I largely adhere to that of Moltmann. When Moltmann says, "An anticipation is not yet a fulfilment. But it

---

8. Cochrane, *Church's Confession Under Hitler*, 237–42.

9. Wells, *Christic Center*, 125.

10. Christ-centeredness is also emphatically Trinitarian because the belief in Jesus Christ is always intertwined with the belief in the Trinitarian God. Moltmann explains it this way: "To confess Jesus as the Lord is at the same time to confess that God who raised him from the dead; and the reverse is also true." Moltmann, *Way of Jesus Christ*, 40.

11. Ibid., xiv.

is already the presence of the future in the conditions of history," he understands *anticipation* as a mediating category between history and eschatology.¹² The understanding of the resurrection of Jesus as the anticipation of the Reign of God in history is different from both "thorough-going eschatology" and "realised eschatology."¹³ In Moltmann's ecological thinking, not only the resurrection of Jesus but also the Spirit that comes through Christ, the gathered congregation of the church, as well as the worship in the Spirit with its liturgy of the Eucharist and baptism are all signs and anticipations of the coming Reign of God, that is, the new creation. In my view, this concept of anticipation—an advance and foretaste of what is yet to come in full scale—not only allows us ever-renewing hope for cosmic redemption in the face of human sin and frustration in the cosmogenetic journey toward the awaited eschaton, but also provides a missionary energy for Christians to transform the relationship between humankind and nature in a dynamic interrelation between the eschaton and present history.¹⁴

METHODOLOGY

The overall methodology of the thesis will be a *mutually critical correlation*¹⁵ of tradition and contemporary situations. Tradition refers to both the Bible and theological concepts developed in the church throughout its history. First, theological tradition will be seen from the perspective of a contemporary situation of unprecedented ecological crisis. In light of this colossal challenge that threatens the total collapse of the ecological system and thus of human existence, the tradition will be re-thought to determine what function it has played in the development of today's ecological crisis and what its contribution could be. Second, I will view the contemporary situation from the perspective of Christian theological tradition. For centuries Christian tradition has not only preserved

---

12. Moltmann, *Church in the Power of the Spirit:*, 193–94.

13. Moltmann calls the "thorough-going eschatology" (represented by, among others, A. Schweitzer) tragic resignation and the "realized eschatology" (represented by, among others, C. H. Dodd) fervent enthusiasm. Ibid., 194.

14. Richard Kearney has proposed, in a non-exclusively religious way, that the potentiality in the eschatological invitation of God brings about transformation in the invitee. Kearney, *God who May Be*, 2, 44–47.

15. Tracy, "Theological Method," 56

the hope and the unquenchable thirst for the Reign of God but also provided a unique source of hope and energy for the renewal of the earth. In addition to a detailed analysis of the sociopolitical situation and scientific and technological innovation, we need vision and fire from above to radically change the stubbornly resistant collective anthropocentrism and ego-centrism to help us re-examine our own false assumptions, and to guide us through this crisis. The Christian Gospel and theological tradition are not resources to lay aside, but something to which we always listen anew and obey. I believe that out of this kind of mutually critical correlation, a truly viable and authentically Christian ecological vision could be born for today.

In clarifying Moltmann's eschatological panentheism, I will use textual analysis. For the purpose of this thesis, I will examine his books and articles, especially paying attention to his doctrine of Trinity, creation, Christology and eschatology. Because I am viewing his theology as a whole inter-related body with a strong eschatological undertone, each book and theme should be seen in light of his overall theological orientation and emphasis, such as his theodicy and anticipatory eschatology, which culminates in his processive eschatological Christology. Therefore, in analyzing Moltmann's eschatological panentheism, I will engage the dialectic of the whole and the parts in order not to lose his overriding themes and undertone. I believe this method will make it possible to highlight the eschatological radicality that Moltmann brings to each theological theme and the christological foundation on which Moltmann places ecological hope, eventually making it possible to bring eschatologically-charged missionary energy to Christian liturgy.

When it comes to the ecological function and possibilities of Christian liturgy, I want to make it clear that this thesis is not a liturgical study but an attempt to open a constructive dialogue with liturgical theology toward an ecological Christian worship. I will start with Moltmann's understanding of Christian worship as an anticipation of the coming Reign of God. On the basis of such theological understanding, I will make suggestions concerning future directions of Reformed liturgy. For that purpose, I will use examples from some of official liturgical documents such as *Baptism, Eucharist, and Ministry* of Faith and Order, "Living Faith" and *The Book of Praise* of the Presbyterian Church in Canada, and the United Church of Canada's *Voices United, More Voices, Celebrate God's Presence,* and *A Song of Faith.*

IMPLICATIONS

First, this thesis will be an articulation of a Reformed perspective on ecological theology. The thesis, however, will not only appeal to Reformed Christians but will also present a Christ-centered ecological hope in a persuasive way to Christians in general.

Second, the thesis will help the church appreciate an eschatological meaning of Christian worship, its Christological-Trinitarian structure and contents, and its missionary energy.

Third, the thesis hopes to open a constructive dialogue between systematic theology and liturgical theology. By bringing to light theological resources for liturgy in Moltmann's eschatological panentheism, pointing out theological principles, and making suggestions regarding the future directions in liturgical development, the thesis hopes to make a meaningful contribution to those who are concerned with harnessing theological insights for developing an ecological liturgy.

Fourth, I hope the thesis will have a positive impact on Korean Christians, many of whom still regard ecological theology as a very liberal theological enterprise, gone too far afield from their understanding of the Gospel. By presenting an ecological hope based on God's faithfulness to God's creation, especially God's faithfulness to the embodied promise of the resurrected body of Jesus Christ, this thesis will help place ecological awareness into the center of Christian hope and ethics.

TOPICAL OUTLINE

In chapter 1 ("Reformed Standards for Envisioning a Christian Ecological Theology"), I will identify standards for a Reformed Christian ecological theology in light of its characteristics and emphases. In order to develop a viable ecological theology, I will also discuss the ecological potential of panentheism as well as its problems in terms of re-envisioning the relationship between humans and the world and between God and the world.

In chapter 2 ("Trinitarian History and Its Glorification in the Cosmos"), I will discuss Moltmann's concept of the Trinitarian process toward eschatological perichoretic union in the Trinitarian community and of the Trinitarian God and the world. For that purpose, I will examine the eschatological orientation, the key concepts of the Trinity and the framework that binds together the whole of Moltmann's theology as a consistent and coherent body. I will also discuss Moltmann's creational concept of God's Sabbath that is pregnant with cosmic hope toward eschatological panentheism.

In chapter 3 ("The Cross and Resurrection as the Foundation for Ecological Redemption"), I will lay bare the Christological concentration and explosive expansion in Moltmann's ecological eschatology. The Christ event of the cross and resurrection is re-interpreted in terms of cosmic tribulation and rebirth toward a new creation in the Trinitarian process toward the cosmic redemption. As a link and preparation for the missionary-ecclesiological concept of worship, I will discuss the role of the Holy Spirit in regard to the Trinitarian-Christological mission.

In chapter 4 ("Christian Worship as Anticipatory Celebration and Missional Sending"), I will present Moltmann's understanding of Christian worship in relation to his eschatological panentheism. I will discuss his understanding of the structure of the worship, the meaning of baptism and Eucharist, and the sending out of the congregation. Special attention will be given to his concept of worship as an anticipatory celebration of eschatological reality and ensuing missionary energy flowing out of it.

In chapter 5 ("Envisioning an Ecological Reformed Worship"), I will first offer some theological principles for an ecological Reformed worship based on Moltmann's eschatological panentheism. Then, I aim to provide some liturgical suggestions regarding future directions of Reformed worship based on Moltmann's theological understanding of worship as an anticipatory celebration of the ecological redemption of the cosmos.

In conclusion, I will summarize how Moltmann's eschatological panentheism represents a viable Reformed approach to ecological theology, as well as how liturgy could nurture ecological spirituality in light of eschatological understanding of Christian worship.

1

Reformed Standards for Envisioning a Viable Christian Ecological Theology

IN THIS CHAPTER, I will articulate the standards necessary for a viable Reformed Christian ecological theology, one both biblically and ecologically informed, both faithful to the Reformed tradition and viable for today's urgent situation. Towards that purpose, I will first survey characteristics of the Reformed tradition. Then I will posit the manner in which today's alleged anti-ecological Christian theology and spirituality can be reformed: by re-envisioning the relationships between humans and the world and God and the world. In this connection, panentheism becomes important. I will discuss the potentials and pitfalls of panentheism as an ecological paradigm. Then I will summarise the discussion to show how Moltmann's theology is able to satisfy the standards for a viable Reformed ecological theology.

STANDARDS FOR A VIABLE REFORMED THEOLOGY

Apologia for a Confessional Approach to Ecological Theology

It might seem odd and inappropriate to start a discussion of an urgently-needed global and far-reaching ecological theology by using characteristics of a particular strand of the Christian tradition. The ecological crisis far surpasses denominational boundaries and is not even Christianity's unique problem. The diagnosis and cure of today's anti-ecological thinking and way of life should be an ecumenical effort and begun with dialogue and cooperation with all those who are concerned with the fate of our common home (*oikoumene*).

That being said, however, we do not have a single ecological theology that can be readily accepted or even tolerated by today's diverse Christian theologies. Although the crisis binds everyone together and the solution to our common problem should be shared, we can do that only when we realize who we are and where we stand theologically and liturgically.

We enter into the Christian faith through a specific faith community and theological perspective, and we continue to be nurtured in that tradition. Until we are conscious of our particular faith background, we cannot begin to appreciate the other voices and perspectives of the larger Christian family, nor can we engage them in meaningful dialogue with a humble spirit. Therefore, without resigning to theological relativism—in the sense that everything is right in its own way and it is useless to pursue a deeper and higher truth through dialogue—we should first endeavor to construct an ecological theology that reflects a way of thinking unique to our own tradition in order to effectively battle our common problem.

Another reason we should start with our own tradition is actual practice. Theology will not save the world if it does not have the power to form and transform people's spirituality. Changing a collective way of thinking and living does not occur only by intellectual and moral persuasion; it involves touching and transforming spirituality. In my view, worship is the occasion during which our whole being, including the deepest part of our soul, can be touched, re-shaped, and nurtured in a certain way and direction.[1] Liturgy is the form with which we participate in this profoundly significant process of divine grace and human response. I am of the opinion that liturgy and theology are so closely interwoven that they shape each other in creative tension.[2] For

---

1. For detailed discussion on this point, see "Worship as the Locus of Moral Formation and Moral Imagination" in chapter 4.

2. Dwight W. Vogel summarizes three possible relationships between theology and liturgy: 1) Liturgy is a source for theological assertions and has priority over them (the patristic period); 2) Theology has priority over liturgy and should judge the adequacy of liturgical formulations (reformation perspectives and feminist critiques); 3) Liturgy and theology affect and ground each other and exist in a creative and symbiotic relationship. Vogel, *Primary Sources of Liturgical Theology*, 11. Among these positions, the

that reason, alien theological ideas cannot easily enter the liturgy of a particular tradition. This is why we should seek to construct a unique ecological theology based on our own tradition and at the same time critically review our liturgical practice in light of theology. Only our own distinctively created ecological theology can be translated into a liturgical mould to shape and nurture an ecological spirituality, for theological approach and liturgy in a certain tradition are closely related and reflect each other.

Therefore, as a Reformed Christian, I will show how my tradition can offer real hope to the current ecumenical dialogue about the ecological crisis and, conversely, what in my tradition can and should be clarified and developed—or re-thought and changed—in light of current ecumenical and scientific understandings. In the following section, I will discuss the characteristics of Reformed theology and principles of Reformed worship as a preparation for establishing appropriate standards for a Reformed ecological theology today.

Characteristics / Principles of a Reformed Theological Approach and Worship

"Reformed" refers to the "churches and theological tradition, as an expression of Christian faith of all times and places, that began with the sixteenth-century Reformation in Zurich, Strasbourg and Geneva."[3] Ulrich Zwingli in Zurich, Martin Bucer in Strasbourg and John Calvin in Geneva were inspired by the theology of Luther and his reformation of the church in Germany, but the differences that surfaced in relation to the understanding of the Lord's Supper became divisive in most cases. As a result, at the end of the sixteenth century, a reformation of their churches had developed under the name *ecclesiae reformatae* that was related to, but also separate from, the Lutheran model.[4]

Lukas Vischer provides the following extensive purview of the characteristics of the Reformed heritage. I summarize here his presentation.[5]

1. *Christus solus*. Jesus Christ is the only and exclusive source of salvation.

---

third one is the closest to my opinion.

3. van der Borght, "Reformed Ecclesiology," 187.
4. Ibid.
5. Vischer, "Reformed Tradition," 26–33.

2. God to be glorified in all things. As a reformulation of the first commandment, it expresses the conviction that our salvation depends entirely on God's initiative.
3. Salvation and Trinitarian thinking. According to the Trinitarian teaching of the creeds of the early church, Reformed teaching affirms that God the Creator of all things is the same God who became human in Jesus Christ and fulfills redemption through the power of the Holy Spirit. Reformed theology places particular emphasis on the saving and healing power of the Holy Spirit.
4. The authority of the Bible. All Reformed confessions converge in stressing the authority of the Scriptures of both the Old and the New Testaments as the source of all decisive knowledge and guide in the life of the church.
5. Confessions of faith. Reformed churches have formulated confessions of faith to affirm and to give account of the truth of the Gospel. They do not possess the same authority as the Bible but are regarded as "subordinate standards."
6. The church. Although there is no way to determine the borderlines of the true church, the church as it exists in history must not be despised. To listen to God's Word and to respond to it, we depend on the community; its message can only be proclaimed through the joint efforts of all. Calvin calls the church the mother who nourishes the faith on the pilgrimage of their life.
7. Prayer and worship. The first response to the proclamation of God's gift of salvation in Jesus Christ is prayer and praise. Worship is primarily a corporate act that consists of prayer, reading scripture, proclamation, and the regular celebration of the Lord's Supper. In their reformation of the church, the Reformers opted for a reformation of the liturgy that removed everything that distracted from these essentials. Images in the churches of the sixteenth century were especially targeted. Today a variety of worship styles has developed within the Reformed churches.
8. Discipleship and discipline. Justified through God's saving grace, we are called to live a life in the church that is inspired by thankfulness and that leads to sanctification. Since unholy lives lead to disintegration of the Christian community, the exercise of discipline against unholy practices and unsound teaching must safe-

guard the integrity of the church.

9. Ministries and church order. Calvin's influential model in Geneva recognised four permanent biblical ministries: pastors to preach the word and administer the sacraments, elders to assist the pastors and exercise discipline, deacons to take care of the poor, and doctors responsible for the pure teaching of the church. In the British Isles, the confrontation with the Episcopal system of the Church of England led to the conviction that the Presbyterian system was the only biblical way to order ministries. The Presbyterians opted for a strong emphasis on the collegial exercise of authority, while the congregationalists stressed the primacy of the local community. In recent decades, a number of unions have taken place in which the polity of the new church combines Presbyterian, Congregationalist and Episcopal elements.

10. The Church—Local and universal. The Reformed marks of the church—pure preaching and appropriate administration of the sacraments—focus on the local community, especially since the mediation of the hierarchical order is notably absent, and the Congregationalists have made this local focus the identity marker of the church. As a result, Reformed churches have developed a tradition of strong participation and responsibility among the members. Most Reformed churches have combined this local accent with a development of common decision-making at regional and national levels through a structure of representation in presbyteries (classes) and synods (assemblies). This system of representative collegial structures has become part of the Reformed heritage.

11. Called to be witnesses of the Gospel. The awareness of missionary calling was absent in the sixteenth century, but has grown strongly, especially through the Revival Movements in the eighteenth and nineteenth centuries, and has become prominent through the practice of mission since the nineteenth century.

12. Truth and unity. Like all churches, Reformed churches face the dilemma of faithfulness to the Gosepl and commitment to the oneness of the Body of Christ. The call for renewal has often put strain on the commitment to the unity of the churches. This dilemma has caused many to abstain from the ecumenical move-

ment, while at the same time the majority have become active in dialogue, collaboration and union.

13. Church and state. The relationship between church and state was a major issue of disagreement in Reformed churches. Basically, the strong emphases on a coherent, constitutionally established internal order of the church militates in favour of autonomy from the state.

14. The witness of the church in society. A preoccupation with the reform of the church has not been understood as being in conflict with the conviction that a call for renewal extends to all aspects of life—the whole society included. At the same time, societal witness is contested among Reformed churches.

15. The church as wandering people. The Reformed churches are on the Way. Facing new situations, they seek to be faithful to God's Word as witnessed to in the Bible and to correct and renew the life of the church accordingly. To point to this readiness Reformed Christians like to cite the formula *ecclesia reformata semper reformanda* (reformed church, always reforming). Despite the danger of abuse of the formula to legitimise change for change's sake, it stands for an important characteristic of Reformed churches—openness to new insights gained through living intercourse with the Bible.

On the other hand, B. A. Gerrish provides a very helpful insight into Reformed theology today. He argues that what defines Reformed tradition today cannot be found in membership to an institution such as World Communion of Reformed Churches, nor historical confessions made in the tradition that is understood as belonging to a particular time and place and thus always carrying *reform*-ability. Reformed theology is not simply preservation and indoctrination of past confessions or an obedient following of the footprints of past Reformed theologians; rather it is a fresh and persistent wrestling with problems on the basis of tradition. Therefore, to achieve *semper reformanda*, Gerrish prefers to point out "habits of mind" in doing theology in the Reformed tradition, rather than relying on fixed tenets, dogmas or past theologians.[6] He proposes five characteristics of the Reformed habit of mind: *deferential* to the Scriptures and our forebears; fundamentally *critical* even to our

---

6. Gerrish, *Reformed Theology for the Third Christian Millennium*, 3–5.

own tradition; *open* to wisdom and truth wherever they are to be found; *practical* with the purpose of edification and sanctification and transformation of society into a mirror of God's glory; *evangelical* in the sense of continually going back to Christ's Gospel.[7]

In addition to these five habits, Reformed theologian Harold G. Wells proposes Christ-centeredness as a non-negotiable theological criterion, with the additions of remaining biblical, contextual and ecclesial as sub-criteria.[8] He argues:

> If we decide theologically that we should be centered somewhere else than in Jesus Christ, if we conclude that our norm for Christian theology is to be found elsewhere, then indeed our language of worship and Christian life must change drastically. A decentered Christ, understood in terms of a reduced Christology, must cease to be so prominent, both in our worship and in our ethical reflection or practice. Hymns of praise to Jesus must cease. Our Trinitarian worship and prayer must be eradicated. We must no longer speak of obeying or following Jesus. To continue as we do in prayer and hymnody, sacrament and preaching, following and obedience would be sheer idolatry![9]

In relation to, and in clarification of, this criterion of Christic center, Wells discusses the creation-centered or cosmocentric theologies. Here, Wells criticizes Sallie McFague's "radicalizations"—her application of the doctrine of incarnation to the universe—as an inadmissible *deradicalization* and a *domestication* of the incarnation.[10] Reading between the lines in his criticism of McFague, one discovers a typically Reformed concern about idolatry.[11] In Wells' view, McFague's understanding of the

---

7. Ibid., 7–8.
8. Wells, *Christic Centre*.
9. Ibid., 125.
10. "Jesus is said to be the incarnation of the divine Wisdom and Word not because he exemplifies, as paradigm, a moral ideal of justice and compassion, and therefore of divine presence, but because his life, death, and resurrection are acknowledged as the eschatological event of salvation. . . . But the created order as such cannot incarnate God, cannot be for us the object of worship, because it does not save or liberate us. In fact, the divine disclosure in Jesus stands in contradiction to the world as we know it, in which typically the strong dominate the weak." Ibid., 129.
11. Kosuke Koyama, a Reformed theologian, also expresses this concern about idolatry in terms of creatures, human culture, technology, and human mental capacity, even the most valuable symbols. His words sound almost like the translation of Latin phrase *soli deo gloria* when he says, "[The holy and impassioned God] does not allow

cosmos as the incarnated body of God, of which Jesus is the paradigm example, divinizes and romanticizes the creation, blessing whatever exists, and tends toward conservative implications. He asks: "If the world is God's body, can we strive to change it radically?"[12] This unintended conservative stance of McFague's panentheism provokes a fear in Wells that it undermines the significance of Jesus Christ, who came as the eschatological transformer of the world as it exists, which for Wells is a "realm of violence, sickness, decay, and death."[13] Wells also points out the close relationship between the theology and worship of the faith community, in that worship expresses theology as well as it helps to form it.[14] This is nowhere clearer than in the Reformed tradition. It is quite true that Reformers were basically reformers of worship and tried to form the whole of life based on the worship of the faith community.[15] In this connection, we can understand how the themes and emphases in Reformed theology are translated and expressed in liturgical form in its worship. For example, the strong emphasis on grace and the covenantal understanding of baptism resulted in an insistence on infant baptism. Also its emphasis on the understanding of humans as images of God—instead of images of God we created—coupled with the Hebrew Scriptures' emphasis on the ban of idols led to the removal and rejection of visible images in worship.[16]

We can detect more examples of the close relation between Reformed theology and Reformed worship. Hughes O. Old informs us of the God-centered, Trinitarian characteristic of Reformed worship. We worship God because God created us and commanded us to worship God, from which comes the first characteristic of Reformed worship: worship according to Scripture.[17] This did not mean literal-

---

himself to be domesticated by us." Koyama, *Mount Fuji and Mount Sinai*, 38–52.

12. Wells, *Christic Centre*, 130.

13. Ibid.

14. Ibid., 122.

15. White, *Protestant Worship*, 65.

16. Old, *Worship*, 128.

17. Regarding this aspect of Reformed tradition, Moltmann is both commended and questioned. On the one hand, it is quite apparent that Moltmann tries to base his argument, or at least to link his theology to, the biblical concepts and narratives. On the other hand, many critics question his use of the Scriptures as arbitrary, unbalanced and exegetically unsound. For example, Farrow criticises Moltmann's use of 1 Cor 15:28 as a panentheistic-sounding proof text without exegetical examination. Farrow, "Review

ism but faithfully following specific directions and examples when they are found in Scripture, and when not, being guided by scriptural principles. Also Reformed worship is in the name of Christ, as the agent of Christ, in the body of Christ. Also Christian worship is Spirit-filled: inspired, empowered, directed, and purified by the Spirit, and bears the fruit of the Spirit by demonstrating a holiness of life. In this way, Reformed worship is a part of God's redemptive work among us through which we are transformed into God's image.[18] Howard L. Rice and James C. Huffstutler also point out six characteristics of Reformed worship: a focus on community; the involvement of the people; simplicity; the combination of Word and sacrament (the Eucharist); the importance of the Psalms; adaptability.[19]

These different sets of principles and characteristics of Reformed theology and worship show the vibrant and diverse aspects of Reformed tradition, but also reveal an enduring inner connection and common emphases of Reformed tradition in its theology and worship. These principles and characteristics, however, tend to converge into the five Latin expressions that signify a Reformed emphases on theology and worship: *sola scriptura* ("by Scripture alone"), *sola fide* ("by faith alone"), *sola gratia* ("by grace alone"), *solus Christus* or *solo Christo* ("Christ alone" or "through Christ alone"), *soli Deo gloria* ("glory to God alone"). These symbols emphatically summarize Reformed tradition and can also be used as a litmus test to assess any theology pretending to be Reformed.

As this thesis aims to assess Moltmann's ecological theology from a Reformed perspective and explores the possibility of liturgical harnessing of his theology in a way that is relevant 1) to today's Christian church in general and the Reformed church in particular and 2) in the context of today's ecological crisis, I formulated the following questions in a way that combines the five *solas* and Gerrish's standards. These questions do not replace the five *solas* but unpack these Reformed symbols in today's context.

Essay," 445, endnote 55. John B. Cobb, Jr. also criticizes Moltmann's uncritical use of Scripture. Cobb, "Reply to Moltmann's 'The Unity of the Triune God,'" 174. Nicholas John Ansell views Moltmann as "imposing" his own theology upon the biblical writings by subordinating apocalyptic material to more universal-redemptive biblical passages. Ansell, "Annihilation of Hell," 374.

18. Old, *Worship*, 1–6.

19. Rice and Huffstutler, *Reformed Worship*, 6–8.

1. Is its worship and theology of the Trinitarian God founded on a christologically-based Gospel, which can be grasped through Scripture (Christ-centeredness)?[20]

2. Does it encompass every aspect of life in the worship and faith-based reflection called theology, aspiring to bring both the whole of individual and communal life under obedience to the Word, with an intended transformation for the glory of God (transformation of the whole of life)?[21]

3. Does it endeavour in a humble spirit to be critical even of itself for the sake of a more appropriate witness in the future and, in the process of such efforts, does it strive to employ the best available knowledge of the time (an ever-reforming and critical mind according to the Scripture)?[22]

---

20. This question combines standards of *soli Deo gloria*, *sola gratia*, *sola scriptura*, and *solus Christus*. This can be identified, in addition to above-cited authors, in "The Theological Declaration of Barmen," in which we can sense a Karl Barth-like uncompromising insistence on the sovereignty of Christ: "Jesus Christ, as he is attested for us in Holy Scripture, is the one Word of God which we have to hear and which we have to trust and obey in life and in death. We reject the false doctrine, as though the church could and would have to acknowledge as a source of its proclamation, apart from and besides this one Word of God, still other events and powers, figures and truths, as God's revelation." Cochrane, *Church's Confession under Hitler*, 239.

21. This question combines standards of *soli Deo gloria*, *sola scriptura*, and *sola fide*. Also in "The Theological Declaration of Barmen," we read, "We reject the false doctrine, as though there were areas of our life in which we would not belong to Jesus Christ, but to other lords—areas in which we would not need justification and sanctification through him." Ibid., 240.

22. This question combines standards of *soli Deo gloria*, *sola scriptura*, and *sola fide*. Also in "The Theological Declaration of Barmen," we read, "'Rather, speaking the truth in love, we are to grow up in every way into him who is the head, into Christ, from whom the whole body [is] joined and knit together.' (Eph. 4:15,16.) The Christian Church is the congregation of the brethren in which Jesus Christ acts presently as the Lord in Word and sacrament through the Holy Spirit. As the church of pardoned sinners, it has to testify in the midst of a sinful world, with its faith as with its obedience, with its message as with its order, that it is solely his property, and that it lives and wants to live solely from his comfort and from his direction in the expectation of his appearance." Ibid.

STANDARDS FOR A VIABLE CHRISTIAN ECOLOGICAL THEOLOGY

Ecological consciousness is fundamentally a realization of the inter-relatedness and inter-dependence of everything. Christianity, however, has long been accused of anthropocentrism, in which humans are viewed as an independent super-species, equipped with a God-endowed privilege of dominion over the rest of the creation. Christian theology cannot sweep this accusation under the carpet, because, even though it did not initiate anthropocentrism as its core tenet,[23] it has at the very least not resisted this kind of human arrogance, indulging itself in a kind of vertical dimension of human salvation, relegating nature to the position of second fiddle. I will introduce Christian anthropocentrism in the form of spirit-matter dualism and the anti-ecological spirituality which is now deeply enshrined in the Western tradition.

Re-envisioning the Relationship between Humans and the World: Christological-Anthropological Questions

"The Great Chain of Being" is a metaphysical concept that reality is a universal, hierarchical structure, in which every being exists along the gradations of the hierarchy, with the One or the Good or God at its peak and a primordial flux and non-being at the bottom. From the apex to the nadir, the spiritual gradually gives way to the material.[24] In this scheme, every existence in the universe has its proper place along a chain vertically extended. Georges Duby explains this emanationist vision in allegedly Dionysius' treatise *Of the Celestial Hierachy—Of the Ecclesiastical Hierarchy* in terms of the effulgence of light:

> At the core of the treatise was one idea: God is light. Every creature stems from that initial, uncreated, creative light. Every creature receives and transmits the divine illumination according to its capacity, that is, according to its rank in the scale of beings, according to the level at which God's intentions situated it hierarchically. The universe, born of an irradiance, was a downward-spilling burst of luminosity, and the light emanating from the primal Being established every created being in its immutable place. . . . And because every object reflected light to a greater or lesser degree, the initial irradiance brought forth from the depths of the shadow, by means of a continuous chain of reflections, a

---

23. This is what Rosemary Radford Ruether asserts in her book, *Liberation Theology*.
24. Santmire, *Travail of Nature*, 45.

contrary movement, a movement of reflection back toward the source of its effulgence. In this way the luminous act of creation brought about of itself a gradual ascension leading backward, step by step, to the invisible and ineffable Being from which all proceeds.[25]

From later antiquity down to the close of the eighteenth century, most philosophers and men of science and, indeed, most educated men, accepted without question a traditional view of the plan and structure of the world.[26] Arthur O. Lovejoy expounds its amazing influence on the thought and history of the Western World. He traces its origin back to Plato, Aristotle and the neoplatonism of Plotinus and points out the three principles—plenitude, continuity, and graduation—which were combined in this conception.[27]

Pertaining to the current discussion on ecology and the relationship between humans and the world is the location of the humans in this Chain. The most important division in this Chain is not between the Creator and the creatures, as in Hebrew thought; instead the great gulf in the hierarchical emanation lies between spirit and matter,[28] which unfortunately runs through the human creature: we are material as body, and are spirit as soul. Thus, this concept could function not only as a self-dividing despising of the human body, but also as a metaphysical background for anthropocentrism, because in the biophysical world, humans are the only creatures that are considered to have spirit.

The Gothic spirit is a Christian spirituality that enshrined the metaphysics of the Great Chain of Being.[29] Erwin Panofsky, in his work

---

25. Duby, *Age of the Cathedrals*, 99f.

26. *Dictionary of History of Ideas*, Volume 1, 326. Online: http://etext.lib.virginia.edu/cgi-local/DHI/dhi.cgi?id=dv1-45.

27. Lovejoy, *Great Chain of Being*, 24–66.

28. Ibid., 47. Also, Santmire, *Nature Reborn*, 54.

29. The fingerprint of the Great Chain of Being cosmology on Christianity appears earlier than the middle age expression of the Gothic spirit in early Christian asceticism. With the growing Hellenisation of Christianity, the ideal became to withdraw from the increasingly secularised world and to deprive the body. Through the works of John Cassian and Augustine of Hippo, who spent part of their lives in the Egyptian desert, the spirituality of the desert fathers deeply affected the spirituality of the following Christian era. Like the Gothic spirit, desert hermits emphasised an ascent to God through periods of self-inflicted purgation of bio-physical and bodily elements that led to unity with the Divine. For a detailed discussion of this movement, see Larkin, "Asceticism," 64–68.

Gothic Architecture and Scholasticism, juxtaposes Gothic cathedrals and the great systematic and synthetic works of the high scholastic era, using Thomas Aquinas' *Summa Theologica* as the supreme example. The High Gothic cathedral sought to embody "the whole of Christian knowledge, theological, moral, natural, and historical, with everything in its place and that which no longer found its place suppressed."[30] Likewise, the High Gothic cathedrals resemble scholastic writing in that both of them are arranged "according to a system of homologous parts and parts of parts."[31] This system of harmonious arrangement, according to Santmire, was the Great Chain of Being.[32]

Unfortunately, of the two ontological movements in the metaphysics of the Great Chain of Being—the *overflowing movement* from God extending to the fullness of created beings below and the *returning movement* of the Many to the One, toward union with the eternal world of the pure spirit above—it was the latter that drove the Gothic spirit.[33] As expressed in the soaring towers of Gothic cathedrals, the spirituality drives the soul to ascend to the spiritual One, high above; salvation lies in the ascent of souls (the Many) to God (the One), leaving behind the world of trees and mountains, animals and oceans to final abrogation. In this spirit, Thomas Aquinas could say there would be no new creation of all things, only spiritual beings.[34]

Although the Great Chain of Being no longer fits with today's cosmology—nor does it fit with the Hebrew concept of God in creation behind the Judaic-Christian world-view—the Gothic spirit still exercises great influence. Santmire finds much evidence of a modified and yet easily identifiable Gothic spirit in today's world.[35] Perhaps it is

---

30. Panofsky, *Gothic Architecture and Scholasticism*, 44–45.

31. Ibid.

32. Santmire, *Nature Reborn*, 78.

33. Ibid., 78–79.

34. Ibid., 79.

35. Teilhard's evolutionary-eschatological vision; twentieth-century Catholic theologian Louis Bouyer's account of cosmology and eschatology in *Cosmos*; the transdenominationally popular language in many forms of Christian worship and prayer that God is high above, an example of which "Lord, I Lift Your Name on High" was the most frequently used American hymn from 1997 to 2005 according to Christian Copyright Licensing International report; many new church buildings that block out nature, especially some mega-churches in the US; some forms of Reformed theology that accentuate the ascension and even the absence of the risen Lord in order to reinforce the theme of his heavenly sovereignty; insidious rise of Gnosticism combined

partly because this spirituality represents an escape to a safer, higher and purer realm, where many suffering people find solace.[36] The chasm between the human spirit and the rest of the bio-physical world defines the typical, incorrigible anthropocentrism seen in so many evangelical Christians today. For many such Christians, the natural world is no kin. The material cosmos exists solely for human consumption; being "spiritual" means ignoring the state and fate of such a mundane world. Behind today's dire ecological state and stubborn insistence on the destructive and aggressive way of human life—in spite of many warning signs—thrives this kind of theological anthropology.

In the introduction of the thesis, I already identified the overcoming of theological anthropocentrism as one of the two common themes of ecological theologies. Moltmann's approach to this problem is Christological-anthropological. By way of a *perichoretic* unity of spirit and body in his anthropology and his emphasis on the bodily nature of Christ's resurrection and its entailing hope for the whole creation, Moltmann tries to provide an antidote for traditional dualism and division in human beings, as well as re-positioning them as a participating member of the creational Sabbath.

---

with Western individualism, and finally the popularity of the "Left Behind" series of novels by Tim LaHaye and Jerry B. Jenkins. Santmire, *Ritualizing Nature*, 98–104.

36. My childhood favorite hymn "I'm Pressing on the Upward Way" that I learned from my mother and countless repetition in public worship in a poor neighborhood in Korea might be a good example:

> 1) I'm pressing on the upward way,
> New heights I'm gaining every day;
> Still praying as I'm onward bound,
> "Lord, plant my feet on higher ground."
> 2) My heart has no desire to stay
> Where doubts arise and fears dismay;
> Tho' some may dwell where these abound,
> My prayer, my aim, is higher ground.
> 3) I want to live above the world,
> Tho' Satan's darts at me are hurled;
> For faith has caught the joyful sound,
> The song of saints on higher ground.
> (Refrain) Lord, lift me up and let me stand
> By faith on Heaven's tableland,
> A higher plane than I have found;
> Lord, plant my feet on higher ground.

Korean Hymnal Society, *Korean-English Hymnal*, number 543.

Re-Envisioning the Relationship between God and the World: A Theological-Christological Question

The traditional Christian emphasis on God's transcendence is another cause of today's ecological problem because the one-sided emphasis on God's transcendence inevitably leads to depriving nature of its sacredness, making it a dull, dumb lump of mere material. Thus, the over-emphasis of this theological concept paved the way for human aggression on, and exploitation of, nature. Today, the relationship between God and the world should be clarified for an ecological re-envisioning of reality. If God is not totally absent in the world but is really present in the physical universe, what is the mode of that presence? Does the affirmation of God's presence in nature cause any theological and political problems? To investigate these questions, I will first examine concepts of deism, pantheism and panentheism.

Deism regards God as the intelligent creator of an independent and law-abiding world. In this regard, deism shares its belief with theism. But unlike theism, deism denies that God providentially guides the creation or intervenes in any way with its course of destiny.[37] Deism posits that God, the powerful Creator of the universe distanced Godself from it and now simply contemplates it. The most well-known image of this kind of God might be the "watchmaker" God, who wound up the universe like a watch and let it run on its own.[38]

Pantheism, on the other hand, posits that all things and beings are modes, attributes, or appearances of one single reality or Being; hence nature and God are believed to be identical.[39] More specifically, pantheism asserts the unity of all reality and the divineness of that unity.[40] Philosophical pantheist Benedict Spinoza asserted that God and nature are but two names for one identical reality.[41] Pantheism has traditionally

---

37. Harvey, *Handbook of Theological Terms*, 66.

38. Hastings, *Encyclopedia of Religion and Ethics*, 541.

39. Harvey, *Handbook of Theological Terms*, 173.

40. "Pantheism parallels naturalism on the first point in that both assert only one reality. But in contrast to naturalism, it calls reality divine. Pantheism is like theism on the second point, for both recognise that the world depends on God. But unlike theism, it does not hold the world's existence to be separate from God's." Ferguson and Wright, *New Dictionary of Theology*, 488.

41. Philosophical pantheist group including Benedict Spinoza and Georg W. F. Hegel often use rationalism, the method of reason unadulterated by sense-data, for knowledge of God. In Christianity, mystics like John Scotus Eriugena, Meister Eckhart,

been rejected by orthodox Christian theologians because it obliterates the distinction between the creator and creation.

Panentheism is the view that attempts to reconcile the insights of pantheism and deism. "If pantheism identifies God and the world taken as a whole, and deism insists that God and the world are separate entities, panentheism argues that the world is included in God's being . . . although the world does not exhaust God's being or creativity."[42] So a panentheistic God penetrates every part of nature, but is nevertheless fully distinct from it. The term "panentheism" comes from the Greek *pan-en-theos*, "all-in-God." While pantheists believe that the universe itself is divine and do not believe in personal or creator gods, panentheists believe God has a mind, created the universe, and cares about each of us personally. In this way, panentheism offers a middle way between deism, which only postulates a god separate from nature, and pantheism, which identifies God with nature.

When Deists remove God from the world, the creation is regarded as profane, wide open to human aggression and exploitation. When pantheists equate God with the world, creation is deemed sacred. But Christianity, especially but not exclusively the Reformed tradition, concerned with idolatry, has refused any hint of pantheism; because of this basic rejection of pantheism, Christianity has emphasized God's transcendence. In this context, panentheism demonstrates its ecological potential, in that it enables the church to affirm both the transcendence and immanence of God *and* allows for the sacredness of the physical world without having to equate nature with God. Moltmann takes a panentheistic stance when he says, "Whereas simple pantheism makes everything a matter of indifference, panentheism is capable of differentiation. Whereas simple pantheism sees merely eternal, divine presence, panentheism is able to discern future transcendence, evolution and intentionality."[43] This point should be one of the standards for a viable

and Jacob Boehme at least border on pantheism. Ibid.

42. Harvey, *Handbook of Theological Terms*, 172.

43. Moltmann, *God in Creation*, 103. Of course it is debatable whether or not Moltmann's position fits a typical panentheistic stance. His eschatological vision of God's eternal Sabbath fits quite nicely fit into panentheism. As to his view on God's creation in the beginning and the continuous creation which will be discussed in detail in chapter 2, however, Moltmann's use of the kabalistic concept *zimsum* postulates a certain kind of distance between God and the creation although God is also connected to the created reality through the Spirit. Thus it is right that Moltmann's conception

Christian ecological theology: recognizing the divine immanence in the creation without compromising divine transcendence.

Another important element for a viable Christian ecological theology would be the power to transform the *status quo*. If theology remains just talk, lacking the real power to change the world, why even bother talking? Therefore, it is crucial to ask whether or not a certain theology has the power to waylay the current destructive path of humankind and how it can lead everyday people to change the world. Hence a third standard for a viable ecological theology: possessing the guiding wisdom and ethical energy needed to transform the *status quo*.

As a summary, I would like to formulate the following three criteria as the touchstone for a viable Christian ecological vision:

1. Does it effectively relocate humankind in its proper place in the ecological web?
2. Does it recognise divine immanence in the creation without compromising divine transcendence?
3. Does it provide the guiding wisdom and ethical energy needed to transform the *status quo*?

THE SUITABILITY OF MOLTMANN'S THEOLOGY AS A VIABLE REFORMED CHRISTIAN ECOLOGICAL THEOLOGY

Until now, I have put forward three criteria for a viable Reformed theology and three criteria for a viable Christian ecological theology. In this section, I would argue Jürgen Moltmann's theology qualifies as a viable Reformed Christian ecological theology.

A Viable Reformed Theology

I would argue Moltmann's theology satisfies the three standards for a viable Reformed theology, namely, 1) Christ-centeredness, 2) the transformation of the whole of life, and 3) an ever-reforming critical mind according to the Scripture.

---

of the relationship between the Creator and the creation is not a simple panentheism but needs nuances and qualifications. Overall, however, with his balanced approach regarding the transcendence and immanence of God vis-à-vis the creation on top of his eschatological panentheistic vision, I argue that we can call him a panentheist.

First, Moltmann's theology is, principally, a *messianic* theology encompassing and re-examining everything including ecclesiology, theology of creation, pneumatology, the doctrine of the Trinity, and ethics under the christologically-determined eschatological light.[44] Furthermore, Moltmann bases ecological hope on the christological Gospel of the cross and resurrection and shows that Christian ecological theology is not a mere excuse or addendum in response to current ecological crisis and attack at Christianity, but something that should arise from the authentic core of Christian faith-hope.[45] But we need to note that, for Moltmann, Christology is always bound up with the doctrines of Trinity and pneumatology, because "to confess Jesus as the Lord is at the same time to confess the God who raised Him from the dead; and the reverse is also true."[46] Thus it is natural that his ecological theology develops in harmony with the doctrines of the Trinity and pneumatology.[47] This inner relation of the doctrine of Trinity, Christology and pneumatology in Moltmann's ecological theology leads to the eschatological glorification in the cosmos and at the consummation of its history, where the whole of creation finally fulfills its created purpose in praising the glory of God. This focal point of Moltmann's eschatological-ecological outlook demonstrates the Reformed habits of mind in ecological theology. Therefore, I would like to submit that Moltmann's theology pursues and is utterly mindful of the Reformed principles of *soli Deo gloria, sola scriptura,* and *solus Christus.*

Secondly, Moltmann's theology also endeavors to incorporate the totality of human life. He takes a critical stance on the separation of nature and history in Christology, maintaining that this tendency of the

---

44. Moltmann himself declares that all his present and future writing could be covered under the umbrella of "messianic theology." *God in Creation*, xvii. To grasp Trinity, Moltmann starts with Jesus Christ as the revealer of the Trinity (more concretely the cross of Jesus). Moltmann, *Trinity and the Kingdom*, 65; the Christology gained there functions as an open one that allows us to perceive of the creation of the world through the Father and of the transfiguration of the world into a new creation through the Holy Spirit who proceeds from the Father through the Son. Ibid., 97. In this way, his Trinitarian framework has the christological gospel as its core.

45. This point will be dealt in detail in chapter 3.

46. Moltmann, *Way of Jesus Christ*, 40.

47. In fact, his doctrine of social Trinity has an intrinsic affinity to ecological thinking about the world. Also, his pneumatology as the "Spirit of life" extends traditional narrow pneumatology to become an ecological doctrine.

last hundred and fifty years has led to the negligence of nature.[48] In his view, nature and history are to be seen as mutually dependent, just like the mind and body. He also expands traditional narrow pneumatology, identifying the eschatological Holy Spirit as a vital and life-giving force and more importantly, as the immanent divine presence in the cosmos.[49] Accordingly, in Moltmann's messianic understanding of the church and worship, it is not merely one functional element in secularised society, but a messianic community that resists a de-humanizing modern society and anticipates the feast of new creation of the cosmos.[50] In this way, Moltmann's ecological theology endeavours to align the whole of individual and communal life along the cosmic hope of new creation. Therefore, I would like to submit that Moltmann's theology pursues and is utterly mindful of the Reformed principles of *soli Deo gloria, sola scriptura,* and *sola fide.*

Thirdly, Moltmann's theology exhibits an ever-reforming critical mind according to the Scripture. He pays much attention to what the Scriptures say, how past Reformers and other important theologians interpreted the Scripture, but he also goes beyond to wrestle with the unprecedented ecological crisis and formulates a viable Christian ecological theology. As a result, he is also critical of the Reformed and, beyond that, the Western traditions. For example, while Moltmann acknowledges the contribution of Karl Barth's theology,[51] he goes beyond Barth to form a more Trinitarian and more fully ecological theology.[52] He also engages himself in a dialogue with evolutionary thinking[53] and modern scientific biological and cosmological development[54] in his attempt to critically incorporate the best available knowledge of our time. He also goes beyond Western tradition to draw on the theologians of the Eastern Church; he finds the unity of the Trinitarian Persons in *peri-*

---

48. *Way of Jesus Christ,* 246–47.
49. *Spirit of Life,* 81–86.
50. *Church in the Power of the Spirit,* 265–67.
51. For example, *Way of Jesus Christ,* 279.
52. Ibid., 188, 230–32.
53. *God in Creation,* 185–214; *Way of Jesus Christ,* 292–305.
54. For Moltmann's use of biological terminology of E. Jantsch, see *God in Creation,* 17. There are, however, criticisms of scientists that Moltmann's uses of these concepts are sometimes at odds with their original implications. Deane-Drummond, *Ecology in Jürgen Moltmann's Theology,* 222.

choresis for the establishment of an ecological Trinitarian doctrine of a social Trinity.[55] Furthermore, Moltmann innovates and re-formulates traditional concepts and doctrines both by recovering the depth and width of understanding to be found in the Scriptures and by employing a contemporary scientific and humanistic understanding of nature and history.[56] Therefore, I would like to submit that Moltmann's theology pursues and is utterly mindful of the Reformed principles of *soli Deo gloria, sola scriptura,* and *sola fide*.[57]

A Viable Christian Ecological Theology

Second, I would argue that Moltmann's eschatological Trinitarian panentheism along with his theology of worship can help us think through these theological-christological-ethical criteria, namely, 1) overcoming undue anthropocentrism, 2) recognising divine immanence without compromising transcendence, and 3) providing guiding wisdom and ethical energy for transformation.

In regard to humankind's place in the ecological web, Moltmann's anthropology posits a *perichoretic* union of spirit and body, and its solidarity and participation in nature. Accordingly, human salvation becomes conceivable only in the context of a cosmic ecological salvation. In this way, Moltmann effectively de-denters humankind from its arrogantly-assumed position of a dominating species to one, albeit a very important one, of the many inter-dependant parts of the creation. Furthermore, his christological-eschatological doctrine of creation re-locates humans alongside the rest of creation toward the eschatological

---

55. *Trinity and the Kingdom*, 150.

56. Among other things, Moltmann tries to incorporate the concepts and discoveries in contemporary evolutionary biology and cosmology into traditional pneumatological concept of fellowship: "The Spirit is the principle of individuation, the principle which differentiates particular 'working sketches' of matter and life on their various levels. Self-assertion and integration, self-preservation and self-transcendence are the two sides of the process in which life evolves. They are not mutual contradictions. They complement one another." *God in Creation*, 100; "This doctrine views creation as a dynamic web of interconnected processes. The Spirit differentiates and binds together." Ibid., 103.

57. The suitability of Moltmann's theology as a viable Reformed theology will be shown in detail in chapters II and III, where his doctrine of creation, Trinity, Christology, and pneumatology are examined. The Reformed characteristics of Moltmann's ecclesiology will be shown in detail in chapter IV, where I present Moltmann's sacramental theology partly in comparison with those of Luther, Zwingli and Calvin.

consummation, in which humanity and the whole creation rejoice in God's indwelling.

Along with other theologian's panentheistic stance, Moltmann's panentheism recognizes divine immanence in the creation without compromising transcendence.[58] *God in Creation*, Moltmann's first book with a consciously ecological purpose, brought a renewed vitality to ecological thought in Christian theology by developing a pneumatology and doctrine of creation in Trinitarian panentheism.[59] In so doing, Moltmann could present a panentheism that is capable of "linking God's immanence in the world with his transcendence in relation to it," by virtue of perichoretic relation of the Persons, resulting in the affirmation of the Spirit both in preservation (immanence) and eschatological transformation (transcendence).[60] Furthermore, *The Spirit of Life* is a pneumatological masterpiece that beautifully fuses the "already" and the "not yet" of the Spirit, and thus validates and expands traditional pneumatological tenets to include future ecological concerns and hopes.[61]

Finally, Moltmann provides guidance, wisdom and ethical energy for the transformation of the *status quo* by placing the Christian community and its worship in the history of the Trinitarian mission. Moltmann's theology of worship—as anticipated celebration and sending-out for participation in Trinitarian mission—provides the

58. For example, Boff, *Cry of the Earth, Cry of the Poor*; McFague, *Body of God*.

59. "It is one-sided to view creation only as the work of 'God's hands' and, as his 'work', something that has simply and solely to be distinguished from God himself. Creation is also the differentiated presence of God the Spirit, the presence of the One in the many" *God in Creation*, 14; "Without the difference between Creator and creature, creation cannot be conceived at all; but this difference is embraced and comprehended by the greater truth which is what the creation narrative really comes down to, because it is the truth from which it springs: the truth that God is all in all. This does not imply a pantheistic dissolution of creation in God; it means the final form which creation is to find in God." Ibid., 89. Also, against the speculative theology of the nineteenth century that adopted a pantheistic idea that identified the Son and the world, Moltmann clarifies: "In order to understand the history of mankind as a history *in* God, the distinction between the world process and the inner-Trinitarian process must be maintained and emphasised." *Trinity and the Kingdom*, 107.

60. "This is the benefit of the Trinitarian doctrine of creation in the Spirit and of the Creator Spirit who indwells creation.... The Spirit preserves and leads living things and their communities beyond themselves." *God in Creation*, 103.

61. In Denis Edwards' words, Moltmann takes both the "sacramental" theology of divine presence and the "prophetic-eschatological" approach looking to the coming of the Spirit which will overturn what is. Edwards, "Ecology and the Holy Spirit," 142–43.

insight as to how we can pneumatologically tap into the eschatological panentheistic vision. The bipolar structure of Christian worship, in terms of christological remembrance and pneumatological hope, allows Christians, on the one hand, to recognise the unredeemed reality of the world and to sigh with the creation in the Spirit, and, on the other hand, to rejoice exuberantly in the vision of the new creation promised and anticipated in the bodily resurrection of Jesus Christ. The tension between these two states—namely, the unredeemed state of the world in all its creaturely limitations and tragedies that Christ took into himself, and the healed, restored, and renewed new creation that Christ's resurrected body presents as an embodied promise—was traditionally enshrined and expressed in static terms of the two natures of Christ in the Chalcedonian Creed. In Moltmann's theology, this tension is now transferred to, and used as fuel for, mission in the world to witness to, and embody that eschatological reality in the context of the unredeemed reality of the here and now.

Panentheism Reconsidered in Reformed Perspective

We have already discussed two sets of standards in our attempt to discern a viable Reformed Christian ecological theology. One final issue before marrying the Reformed tradition with Christian ecological theology is this question: how can panentheism differentiate the presence of God in the physical world and the presence of God in Jesus Christ?

Christian doctrine states that Jesus is the incarnate Word and Wisdom of God; His life, death and resurrection is believed to be the eschatological event of salvation. How can we sustain this truth claim if we affirm all reality is the embodiment of God, imbued with the Holy Spirit? Thus, the question becomes important: how can we effectively differentiate the modes of presence of divinity in Jesus and in the universe in general?

As we have already seen in Wells' criticism of McFague's panentheism, a typically-Reformed concern arises regarding whether or not panentheism divinises nature, and by extension, the *status quo* in general. If God is already in the physical and natural world, then they should be regarded as sacred, as an *embodiment* of divinity and a demonstration of divine will. This line of thought could easily lead to a political conservatism that could function as a thoroughfare on, instead of an obstacle to, our current path of destruction.

Moltmann's eschatological theology effectively overcomes this problem with his reservation of complete perichoretic indwelling to the eschatological Sabbath or the new creation. First of all, for Moltmann, Christian panentheism was conceived to emphasise the *otherness* of the creation. The Father loves the Son eternally and the Son returns the Father's love. But this is intra-Trinitarian love, the love of "like for like," not love of the Other, which God desired out of God's divine essence of love. According to Moltmann, in this paradigm, "creation is a fruit of God's longing for 'his Other' and for that Other's free response to the divine love."[62] Second, Moltmann proposed an *eschatological panentheistic* vision of God's Sabbath, where God makes the creation God's home, fulfilling it with God's perichoretic love. That is why Moltmann says that the truth that "God is all in all" is "the *final* form which creation is to find in God."[63] By this eschatological panentheistic vision, Moltmann affirms and upholds the intrinsic value of nature, while avoiding divinising the *current* natural order as an eternal divine will. The world is the promise and anticipation of what is to come and is going through the preparation process, in which the Spirit is already pervasively present in, and re-moulding, the cosmos, and the Son is walking through time to lead the creation to its culminating point. Here, Moltmann's processive Christology and his creative pneumatology set the tone for the tension-filled Trinitarian mission into the cosmos, in which nature and history are inseparably combined. With his processive Christology based on a strong awareness of theodicy, along with his emphasis on the bodily nature of Christ's resurrection, his *eschatological* panentheism effectively avoids the divinisation of nature.

Of course Moltmann's presentation of his eschatological panentheism is not without problems and he is sometimes rightly and sometimes unfairly criticized for them. His understanding of the current state of the creation as the winter of creation combined with his understanding of sin, for example, invited harsh criticisms from Reformed scholars including Douglas J. Schuurmann and Steven Bouma-Prediger as confounding creation and fall.[64] John Cobb Jr. finds Moltmann's use of Scripture uncritical and arbitrary.[65] Others find Moltmann's concepts loose and

---

62. *Trinity and the Kingdom*, 106.

63. *God in Creation*, 89. The emphasis is mine.

64. Schuurmann, "Creation, Eschaton, and Ethics"; Bouma-Prediger, *Greening of Theology*.

65. Cobb, "Reply to Jürgen Moltmann's 'The Unity of the Triune God,'" 173–77.

often contradictory, perhaps due to his explanatory approach rather than systematic approach. Although they hastily add that the richness, boldness, and seminality of his thoughts outweigh his shortcomings, it is true that Moltmann's arguments are not always clear and easy to follow.[66] In the following chapters, I will deal with these criticisms and evaluate to what degree they are legitimate. With all his problems and some unresolved questions, however, Moltmann's ecological theology emerges as a promising theological vision for the Christian community, especially for those in the Reformed tradition.

CONCLUSION

In this chapter, I set out to present standards for a viable Reformed theology from characteristics and principles of a Reformed theological approach and worship, namely, Christ-centeredness, transformation of the whole of life, and ever-reforming critical mind according to the Scripture.

I also defined the critical tasks of Christian ecological theology in terms of re-envisioning the relationship between humans and the world and that between God and the world. I re-formulated standards for a viable Christian ecological theology: 1) Does it effectively relocate humankind in its proper place in the ecological web? 2) Does it recognize divine immanence in the creation without compromising divine transcendence? 3) Does it provide the guiding wisdom and ethical energy needed to transform the *status quo*?

Then I have shown that Moltmann's eschatological panentheism satisfies both sets of standards with a brief presentation of the characteristics, the orientations and inner-relations of his theology. Then I emphasized the special relevance of an eschatological panentheism for today's Christian ecological theology from the Reformed perspective.

In the following chapters, I will show in detail how Moltmann's ecological theology satisfies the criteria of a Reformed Christian ecological theology.

---

66. Among others, Farrow, "Review Essay: In the End is the Beginning," 427; Bouma-Prediger, *Greening of Theology*, 103–6.

2

Trinitarian History and Its Glorification in the Cosmos

THE TRINITARIAN FRAMEWORK OF MOLTMANN'S ECOLOGICAL THEOLOGY

Ecology and Trinitarian Thinking

Ecological Cosmology and Ecology

COSMOLOGY IS A RELIGIOUS / cultural / scientific expression of a given time to describe how God relates to the world and everything in it, and/or to postulate the origins of the Universe. Therefore, every cosmology at a given time produces "a global image of the world to indicate its connection with the divine dimension."[1] Leonardo Boff talks about three familiar types of cosmologies in the Western tradition:

1) In the cosmology of antiquity, the world is a hierarchical, sacred, and unchangeable whole, with what Santmire calls "The Great Chain of Being" as its characteristic metaphor. In the ladder, God the Creator is at the top and humans occupy a relatively important position. This model can be called theocentric cosmology.[2] It is also noteworthy that this cosmology is static.

---

1. Boff, *Ecology and Liberation*, 62.
2. Ibid.

2) The modern cosmology was developed on the basis of the physics of Newton, Copernicus and Galileo, as well as the Cartesian scientific method. Unlike the first model in which everything was connected in a hierarchical way, this model presents a dualistic world, where matter and spirit are disconnected. Leaving the world of spirit to philosophy and theology, science delves into the inactive, dead realm of matter, which can best be described as a machine that God the great Architect built and put in motion with unchangeable laws. This material world is waiting to be discovered and subjugated by humans (beings with spirit), with science as the weapon they employ to extract the secret. This cosmology is anthropocentric.[3]

3) Ecological cosmology is developed from Einstein's theory of relativity; from the quantum physics of Bohr; from the indeterminacy principle of Heisenberg; from the theoretical physics of Prigogine and Stengers; and from the contributions of depth psychology, transpersonal psychology, biogenetics, cybernetics, deep ecology and systems theory and thinking. In some aspects, this new cosmology resembles the first ancient model, in that it presents a unified inter-related universe. But unlike the first model, the new cosmology proposes a vision of an unhierarchical world. Also unlike the second model of a dichotomised world, the new cosmology upholds a fundamentally connected, organic, and holistic vision of the world. The dichotomies of matter and spirit, and thus body and soul, are replaced by the language of life and energy in a profoundly interrelated whole.[4] As a result, the human being is no longer seen as over and above reality—that is, the great cosmic community that is always emerging and changing—but as a participant in a whole with a special functional but non-hierarchical role.[5]

In ecological cosmology, the fundamental connectedness of everything has a pivotal importance for an ecological perspective. More than any other science, ecology confronts nature as an organic, differentiated, and single whole.[6] With this perspective, ecology does not represent a specialized, fragmented knowledge but is defined as a "knowledge of

---

3. Ibid., 62–63.

4. Therefore, the new cosmology marks a break from the theistic worldview of the second model, in which the absolute God is separated from the world, unaffected and unchanging. Boff, *Cry of the Earth, Cry of the Poor*, 141.

5. Boff, *Ecology and Liberation*, 63–64; idem, *Cry of the Earth, Cry of the Poor*, xii.

6. Boff, *Ecology and Liberation*, 45.

interrelated knowledges," a "science of the symphony of life."[7] Boff takes up the definition of Ernst Haeckel that "ecology is the study of the interrelationship of all living and nonliving systems among themselves and with their environment."[8] Therefore, "What is specific about ecological discourses lies not in the study of one pole or the other but in the interaction and interrelationship between them."[9] This interaction and interrelationship goes beyond the reality of a given moment. With its peculiar transversality, ecology "relates laterally (ecological community), frontward (future), backward (past), and inwardly (complexity) all experiences."[10] In this sense, ecology is indeed the science of the symphony of life.

Moltmann's ecological theology adds a theological dimension to this all-comprehensive ecological web of thinking: the Trinitarian God as the environment of the cosmogenetic reality, the source and ground for it, and the guiding force to its *telos* through an intimate interaction with it. I believe this is what he means when he said:

> For centuries, men and women have tried to understand God's creation as nature, so that they can exploit it in accordance with the laws science has discovered. Today the essential point is to understand this knowable, controllable and usable nature as God's creation, and to learn to respect it as such. The limited sphere of reality which we call "nature" must be lifted into the totality of being which is termed "God's creation."[11]

Ecology and Trinity

As is obvious from the discussion above, the new ecological cosmology and ecology as the science of the symphony of life have innate pro-Trinitarian inclinations. First, unlike the cosmology of antiquity, the ecological cosmology sees diversity in the constitution of the cosmos and it senses movement and change to attain a deeper communion. This perception can be clearly linked to the Trinity, a community of differentiated Persons, ever pursuing deeper communion.

---

7. Boff, *Cry of the Earth, Cry of the Poor*, 3.
8. Ibid.
9. Ibid.
10. Ibid., 4.
11. Moltmann, *God in Creation*, 21.

Second, unlike the modern cosmology that posits a deistic dichotomy of matter and spirit to make the world a machine-like, inanimate chunk of matter, the new ecological cosmology presents the universe as multi-layered and containing many aspects within a unified reality. Without invoking a foreign origin, this encompasses everything, including the cosmos' interiority and its inclination towards a higher degree of complexity, of which human beings and their consciousness represent the climactic example.[12] This leans toward Christian panentheism, in which God permeates the world without totally identifying with it.

Third, the concept of a cosmic community, constituted of equal beings that depend on each other in a profound way by reason of an inconceivably complicated web of inter-relations, is also consonant with the Trinity, a community of three Persons that are equal in dignity and power, giving and residing in each perichoretically.

Trinitarian Suffering and Ecological Hope in Moltmann

For Moltmann, his doctrine of the Trinity is an ecological doctrine. Even before he takes on the task of presenting an ecological doctrine of creation, Moltmann was aware of the ecological themes latent in his social doctrine of the Trinity.[13] Before delving into Moltmann's doctrine of the Trinity, however, there is one point to look into regarding the relationship between his doctrine of the Trinity and ecological thinking. Does Moltmann's doctrine of the Trinity take seriously enough the ecological suffering of the creation that includes both humans and their fellow creatures? Does Moltmann's ecological doctrine of God really address the pain and hope of the suffering creation and relate it to the innermost essence of God's being, instead of the occasional intrusive, condescending "fixing" operation that traditional theism taught us as the doctrine of providence? These questions are important for the following reasons: 1) to identify the inner relation between Moltmann's earlier trilogy and his later ecological writings; 2) to examine if his ecological theology as a whole is Christ-centered and thus can appeal to Christians in general and especially to the Reformed Christians; and 3) to determine whether

---

12. Thomas Berry and Brian Swimme, *Universe Story*, 76.

13. "By taking up panentheistic ideas from the Jewish and the Christian traditions, we shall try to think ecologically about God, man and the world in their relationships and indwellings. In this way it is not merely the Christian doctrine of the Trinity that we are trying to work out anew; our aim is to develop and practice Trinitarian *thinking* as well." *Trinity and the Kingdom*, 19–20. The emphasis is the author's.

his ecological theology is an escapist daydream in the face of an urgent crisis or whether it can form a strong ethical basis to overcome the crisis.

First, his doctrine of creation has the suffering of God at its heart.[14] For Moltmann, even the idea of *creatio ex nihilo* portrays the suffering of God. Using the concept of *zimsum*,[15] a *cruciform* of Moltmann's doctrine of creation,[16] he demonstrates that creation out of nothing was preceded by the withdrawal-within, limitation, inversion, self-negation, *kenosis*, and suffering of God.[17] Moltmann says:

> Creation is preceded by this self-movement on God's part, a movement which allows creation the space for its own being. He 'creates' the preconditions for the existence of his creation by withdrawing his presence and his power.... It is the affirmative force of God's self-negation which becomes the creative force in creation and salvation.[18]

In line with this kind of thinking about the creation, one expects Moltmann to redefine God's power in terms of suffering and love. "The sole omnipotence which God possesses is the almighty power of suffering love.... This is the essence of divine sovereignty."[19] Here, God's almighty power is redefined as the suffering love of God. This suffering and humiliation of God is continued in covenantal history, the Babylonian exile, martyrdom, and reaches its consummation at the cross of Jesus.

---

14. This point will be dealt with in more detail in section 2.1.1. "Doctrine of Creation Reconsidered in Light of Christian Trinitarianism" in this chapter.

15. Zimsum means "concentration and contraction, and signifies a withdrawing of oneself into oneself." *God in Creation*, 87. On its origin and Moltmann's use of this concept is discussed in the section "*Zimsum*: Divine Self-Contraction".

16. Deane-Drummond, *Ecology in Jürgen Moltmann's Theology*, 126.

17. Ibid., 86–93. On this point, one could argue that withdrawal and limitation are not really suffering but only a potential for suffering. I would like to argue that the relational nature of God in creation/conception of the cosmos already contains not only the potential for future suffering as a direct sin and revolt of the created beings, but also internal complication in the Creator in adjusting to and being bound in a covenantal relationship with the creation/baby. In the process of preparing herself for the growing existence of the baby, a mother suffers from the conceding of a space in her and the ensuing hormonal changes. In this regard, I agree with Deane-Drummond that *zimsum* is a cruciform of the doctrine of creation, in that it implies conceding, giving up, imparting of self for the other.

18. *Trinity and the Kingdom*, 87.

19. Ibid., 31.

Furthermore, the suffering love is attributed to the parallel history of the Holy Spirit both in creation and, as God's immanent presence in the world, throughout the history of created beings. Moltmann says: "If God commits himself to his limited creation, and if he himself dwells in it as 'the giver of life,' this presupposes a self-limitation, a self-humiliation and a self-surrender of the Spirit. The history of suffering creation, which is subject to transience, then brings with it a history of suffering by the Spirit who dwells in creation."[20] In this perspective, Moltmann could say that "human suffering (*sympatheia*) is a reflection of, and participation in, God's suffering (*pathos*) which is in fact suffering love."[21]

What is important about this suffering of God in relation to ecology is that the loving suffering of God before and in human history is already a promise and anticipation of the creation's eschatological glory. Moltmann puts it this way: "These accommodations of God to the limitations of human history at the same time contain anticipations of his future indwelling in his whole creation, when in the end all lands will be full of his glory."[22] The pouring out of the Holy Spirit on all flesh (Joel 2:28), for example, is a radical limitation on the part of God with all the bodily character of the Spirit; at the same time, however, it becomes a guarantee for the full eschatological indwelling of God in the whole creation.[23] Therefore, a loving God is a suffering God, a self-giving and self-humiliating God. This loving nature of God begins to be seen in the creation and the subsequent history, both christologically and pneumatologically, culminating at the Trinitarian event of the cross. God's suffering, however, forms the basis of ecological hope for the creation. In this way, we could determine that, for Moltmann, the suffering of God and human beings is deeply intertwined with ecological hope.

Characteristics of Moltmann's Doctrine of the Trinity

Moltmann's doctrine of the Trinity was, from the beginning, never a doctrine of God separate from God's creation; the doctrine of God is, for Moltmann, always a doctrine of God's dealings with the world. This stance renders the strong separation of an immanent Trinity and an eco-

---

20. *God in Creation*, 102.
21. *Crucified God*, 267–78.
22. Ibid., 273.
23. *God in Creation*, 67.

nomic Trinity problematic. It also tends to make Moltmann's doctrine of the Trinity largely a doctrine on the Holy Spirit (in close relation to his Christology, of course), for the Spirit is the efficacy of divine work.[24]

Moltmann's Trinitarian thinking is already expressed in *The Crucified God* and *The Church in the Power of the Spirit*, although the pneumatology in the latter is mainly in relation to the nature and task of the church. Moltmann himself felt the need to put forth the Trinitarian framework of his theology that has already been at work in his trilogy. If it is true, as Moltmann maintained, that Christ is the revealer of God, he needed to provide a more inclusive symbol for God's promise seen from the resurrection (*The Theology of Hope*) and God's suffering seen from the cross (*The Crucified God*). He also needed to put those ideas in the context of a story, a narrative that gives an intelligible frame in understanding not only the meaning of the Christ event but the beginning (creation) and the end (*eschaton*) of the world. The all-important Christ event of the cross / resurrection and the insight into cosmos-encompassing Trinity from that event needed to be expressed in a more systematic way in *The Trinity and the Kingdom*.

A full-fledged Moltmannian doctrine of the Trinity is developed on the basis of the Orthodox Church's social Trinity. At the same time, however, Moltmann never loses sight of the all-important christological event and establishes a doctrine of the Trinity built on the eschatological-christological understanding of the Gospel. This is consistent with a direction he takes in his earlier trilogy, where he foresees the eschatological future on the basis of Christ's resurrection and looks back to a grim historical reality through the lens of Christ's crucifixion and identifies the church in the Spirit as that which arises out of the Christ event of the cross and the resurrection.

Social Doctrine of the Trinity

Moltmann's doctrine of the Trinity is primarily influenced by the Eastern Church. Moltmann feels the traditional Western approach to this doctrine is fraught with difficulties. We cannot turn to the earlier Trinity of substance because of its presupposed old cosmology. Neither can we return to the more modern "subject" Trinity because of its increasing lack of persuasiveness in light of the new, relativistic theories about the world. In fact, anthropocentric behaviour increasingly tends to be

---

24. Ibid., 9, 13.

explained in terms of social patterns and an increasing awareness of a fundamental interdependence of everything in the world rather than isolated and disembodied self.[25] Moltmann here turns from dead-ended Western Trinitarian thinking that always starts with God's *unity* to the Eastern thinking in which one starts "with the trinity of the Persons" and then goes on "to ask about the unity."[26] With this approach, Moltmann arrives at a "concept of the divine unity as the union of the tri-unity."[27] Here, the concept of *perichoresis* becomes very important to understand the union of three divine Persons.[28]

Christ-Centered Doctrine of the Trinity

Second, in spite of the influence of the Eastern Church his Trinitarian doctrine is strongly christological. For Moltmann, "God cannot be comprehended without Christ, and Christ cannot be understood without God."[29] Due to his strong concern for theodicy, he identifies the Trinity from the cross of Christ and starts his reflection on the Trinity with the history of the Son.[30] He wants to ask what it means to do theology after Auschwitz. He does not want to construct a concept of God that is separated from the convolutions of history, from the suffering of the people and the creation. In fact, he identifies the task of theology as "draw[ing] the hoped-for future already into the misery of the present and us[ing] it in practical initiatives for overcoming this misery."[31] Therefore, in constructing a doctrine of God, Moltmann's concern is not an abstract question of God's existence but "the *rule* of this God in heaven and on earth."[32]

---

25. *Trinity and the Kingdom*, 18–19; Boff, *Cry of the Earth, Cry of the Poor*, 141.

26. *Trinity and the Kingdom*, 19.

27. Ibid; also, "the unity of the Trinity cannot be a monadic unity. The unity of the divine tri-unity lies in the *union* of the Father, the Son and the Spirit, not in their numerical unity. It lies in their *fellowship*, not in the identity of a single subject." Ibid., 95. Emphases are the author's.

28. This point will be elaborated below in the section titled "*Perichoresis* in Trinity."

29. Ibid., 132.

30. "In order to grasp the Trinity in the biblical history, let us begin with the history of Jesus, the Son, for he is the revealer of the Trinity. It is in his historical and eschatological history that we can perceive the differences, the relationships and the unity of the Father, the Son and the Spirit." *Trinity and the Kingdom*, 65.

31. Moltmann, *Religion, Revolution, and the Future*, 140.

32. *Trinity and the Kingdom*, 191.

This stance, already prominent in his early trilogy, leads Moltmann to a strongly Christology-based Trinitarianism.

Inseparable Mutual Relationship between the Immanent and Economic Trinity

Third, Moltmann distinguishes the immanent Trinity and economic Trinity but does not separate them. For him, "the relationship of the triune God to himself and the relationship of the triune God to his world is not to be understood as a one-way relationship—the relation of image to reflection, idea to appearance, essence to manifestation—but as a mutual one."[33] Although Moltmann admits that the divine relationship to the world is primarily determined by the inner relationship,[34] as is evident in his argument that the creation is the fruit of God's loving nature, he asserts that the economic Trinity has a *retroactive* effect on the immanent Trinity.[35] More concretely, this retroactive effect takes place in the christological / pneumatological event of the cross and the responsive love of the creation in union with the Son in the Spirit. In this connection, Moltmann approvingly introduces C. E. Rolt's idea: "What Christ, the incarnate God, did in time, God, the heavenly Father, does and must do in eternity. . . . For 'the mystery of the cross' is a mystery which lies at the center of God's eternal being."[36] Because of this kind of intimate and non-separable relationship between the immanent and economic Trinities, Moltmann declares: "The pain of the cross determines the inner life of the triune God from eternity to eternity. . . . The joy of responsive love in glorification through the Spirit determines the inner life of the triune God from eternity to eternity too."[37] This third characteristic is the logical connection between the former two characteristics.

---

33. Ibid., 160–61.

34. "Before God issues creatively out of himself, he acts inwardly on himself, resolving for himself, committing himself, determining himself." *God in Creation*, 86; "His creative activity outwards is preceded by this humble divine self-restriction. In this sense God's self-humiliation does not begin merely with creation, inasmuch as God commits himself to this world: it begins beforehand, and is the presupposition that makes creation possible." Ibid., 88.

35. *Trinity and the Kingdom*, 161.

36. Ibid., 31.

37. Ibid.

The Concept of the Trinitarian History of God

Christological History and Trinitarian History

As is already evident from *The Crucified God*, Moltmann regards the doctrine of the Trinity as a formal expression of the crucifixion,[38] and, conversely, in light of the doctrine of the Trinity, he believes, we can understand Christ's history as the history of God.[39] For Moltmann, the Trinitarian history of God is grasped from the history of Christ: if one looks backward and asks where the history comes from, one glimpses the sending of Christ by His Father in the Spirit into the world; if one looks forward and asks where this history of Christ goes, one envisions the implication of Christ's resurrection from the dead to the Father in terms of new creation of *ta panta*. In this Christ-centered Trinitarian thinking that looks backward and forward pneumatologically, Moltmann finds a symmetrically overarching Trinitarian history in which vividly contrasting correspondences flow out of Christ's history into His origin and His future:

> His sending points to his origin with the Father. His resurrection points to his future with the Father. His messianic sending in the world corresponds to his eschatological gathering of the world. His pre-existent origin corresponds to his eschatological future. His becoming human in time corresponds to his becoming divine (*theosis*) in eternity. His surrender to death on the cross corresponds to his exaltation to the right hand of God. His passion corresponds to his transfiguration and his descent into hell to his ascension into heaven.[40]

Moltmann protests that in the classical doctrine of the Trinity, the sending was always at the center—the sending of Jesus, the incarnation of the Son of God, His history of suffering, and the ultimate sending, His

---

38. "The material principle of the doctrine of the Trinity is the cross of Christ. The formal principle of knowledge of the cross is the doctrine of the Trinity." *God in Creation*, 241. Similarly, Moltmann argues that "The experiential content of Trinitarian concept of God is in fact the cross of Christ on Golgotha. The conceptual framework for understanding this history of Christ as the history of God is the doctrine of the Trinity." "Trinitarian History of God," 633.

39. "We interpreted salvation history as 'the history of the Son' of God, Jesus Christ. We understood this history as the Trinitarian history of God in the concurrent and joint workings of the three subjects, Father, Son and Spirit; and we interpreted it as the history of God's Trinitarian relationships of fellowship." *Trinity and the Kingdom*, 156.

40. Ibid., 638.

surrender to death on the cross—whereas the eschatological statements about the history of Christ, His resurrection, His exaltation, His transfiguration, and His handing over the Kingdom to the Father faded into the background.[41] The one-sided emphasis on the sending and ignoring of the eschatological process in the Trinitarian history represent a distortion in Christian theology that is partly responsible for the current deficiency of ecological hope and its entailing practice in the church.

In this line of thinking, Moltmann develops the concept of a Trinitarian history of God in accordance with the two-fold understanding of the history of Christ: the "Trinity in sending and origin" and the "Trinity in glorification."[42]

Trinity in Sending

Moltmann notes that sending is the epitome of the relations among the divine Persons and their common relation to the world based on the perception of the history of Jesus Christ. In the Western tradition, where the sending constitutes the inner Trinitarian relationships, the Father is always only the sending one, while the Son and the Spirit are sent, although the Son is also seen as a sender (*filioque*).

The experiential basis of the perception of the Trinity in sending is, of course, the peculiar history of Jesus Christ. But Moltmann proceeds to reason that if "God appears as the sending Father and the sent Son in history, so must he first and foremost be in himself" because the sending of Jesus must correspond to God and not merely an historical accident.[43] Here, Moltmann reasons back from the perception of the historical Jesus as the sent one, to His eternal sending and His *generatio* from the Father, and from the experience of the Spirit as the sent one to His/Her eternal sending and His/Her *spiratio* from the Father.

Moltmann's inference from historical to theological, or the grounding of the sending in "Trinity in origin," is vital because the Trinity can be shown as *open* toward the creation from eternity. The concept of "Trinity in origin" drawn from "Trinity in sending," by the logic of correspondence, relates the whole history of the world to the very nature and being of the Trinitarian God by showing that it was expecting, invit-

---

41. Ibid., 639.

42. Ibid., 635–43.

43. Ibid., 636. This linking of economic Trinity and immanent Trinity is more fully developed in his later work, *Trinity and the Kingdom*, 159–61.

ing, and intending to embrace the world in its fundamental orientation. In the sending of the Son and the Holy Spirit, the Trinity really opens up Godself for the experience of history. Therefore, in the concept of "Trinity in sending" and "Trinity in origin" drawn from it, the world is taken up into the being and action of the Trinity. In Moltmann's words,

> The Trinity is open for its own sending. It is thereby open for humanity and for the whole created, non-divine world. The sending of the Son for the salvation of the world and the sending of the Spirit for the uniting of the world with the Son and with the Father can therefore also be designated, in a summary way, the *love of God* which proceeds out of itself.[44]

The concept of Trinity in origin with the concept of Trinity in sending as its basis, however, already indicates that it is not a static concept of the Trinity. Because it shows the openness of the Trinity toward the world and its intention to embrace the world, we must expect a process, a history, and a future, in which the Trinity gathers, transforms, unites, and glorifies the world as the other side of the Christ event.

Trinity in Glorification

The other side of sending is glorification which coincides with the completion of the mission of Christ in the Holy Spirit. Whereas, in the Trinity in origin, everything starts with the Father and then the Son—and only eventually the Holy Spirit who proceeds from the Father and through the Son—in the framework of the eschatological unity of God, it is the Holy Spirit that is spoken of first, then the Son and finally, the Father. The Holy Spirit is the One who glorifies the Son as well as the Father. The Son is glorified by the Spirit and glorifies the Father.

The mission of Christ is fulfilled in the glorification of the believers and of the creation. Precisely in and with this, Christ comes to His own glorification. But the Holy Spirit is the power who makes certain and effects this glorification of the believing community of Christ. The Holy Spirit glorifies Christ in the believers and *joins* them with the new humanity of the resurrected one.

It is important to note here that the Holy Spirit glorifies the Son and the Father with something other than the "Trinity in origin." Therefore the "Trinity in origin" that has been opened up by sending the Son and

---

44. Ibid., 637.

the Spirit has been changed by participation in the history of the creation; God has gathered into Godself the experience of history.

Another important aspect of this glorification of the Trinity is that it is also the unification process of the Trinity. For Moltmann, the unity of God is an eschatological reality[45]; he even prefers to use the phrase *unifying* (*Vereinigung*) instead of unity (*Einheit*).[46] Here, Moltmann develops the idea of "unifying of God" out of the Jewish mystical doctrine of *Shekina*, a part of God cut off from Godself given to Israel to suffer with its sufferings. If the *Shekina* shares in the people's sufferings and the agony of exile and joins their wanderings, the redemption of God's Self is also bound with the redemption of Israel.[47] In this context, Moltmann cites Peter Kuhn's rabbinic study: "Israel knows that it will be redeemed because God will indeed redeem God's Self and thereby also God's people."[48] Thus, Rosenzweig interprets acknowledging God's unity in *Shema Israel* as uniting God.[49] Moltmann asks:

> Is not what is here, according to Rosenzweig, entrusted to Israel, in an analogous way entrusted by Christian thinking to the Holy Spirit, who through believers 'unites' God by glorifying him? Does not God's separatation from himself in order to suffer with his people correspond on another level to the separation of God the Father from his Son in the cross, in order that he might suffer with the Godforsakenness of the goldless and so vicarously abolish it?[50]

---

45. The most extreme manifestation of God's separation in the economic Trinity is, of course, the event of the cross at Golgotha. "The Trinitarian self-distinction of God in the death of the Son on the cross is so deep and so broad that all those lost and abandoned will find a place in God." Lapide, Moltmann, and Swindler, *Jewish Monotheism and Christian Trinitarian Doctrine*, 53.

46. "In the eschatological considerations of historical thinking, the idea of the unity of God has a high soteriological content. For that reason it would be better to talk in this respect about the 'union' of God." *The Church in the Power of the Spirit*, 61; Moltmann, *Kirche in der Kraft des Geistes*, 77. The word "union" in English translation does not fully capture "vereinigung"; it is rather *unifying*.

47. The concept of *Shekinah* and its significance for Moltmann's doctrine of creation will be considered in more detail later in th section titled "*Zimsum*: Divine Self-Contraction" in this chapter.

48. Kuhn, *Gottes Selbsterniedrigung in der Theologie der Rabbinen*, 89–90. Cited in Lapide, Moltmann, and Swindler, *Jewish Monotheism and Christian Trinitarian Doctrine*, 50.

49. *Church in the Power of the Spirit*, 61.

50. Ibid.

In this analogical thinking, Moltmann arrives at the conclusion that the eschatological process of the glorification by the Holy Spirit through believers and the whole creation is also a process of a Trinitarian unifying process, which is not a return to the worldless unity of the Trinity in origin, but a Trinitarian *koinonia* that includes the cosmos united to the Son by the Spirit. In fact, because God will be Godself again only when this process of unification is complete, "God and the world are then involved in a common redemptive process."[51] Thus, "The deliverance of the world from its contradiction is nothing less than God's deliverance of himself from the contradiction of his world."[52] In this sense, Moltmann declares: "The unity of the triune God is the goal of the uniting of man and creation with the Father and the Son in the Spirit."[53]

By way of summary, the relationship among Moltmann's concepts such as Christ's history with the cross / resurrection as its heart, the Trinity in origin, and the Trinity in eschatological union can be expressed in the following way:

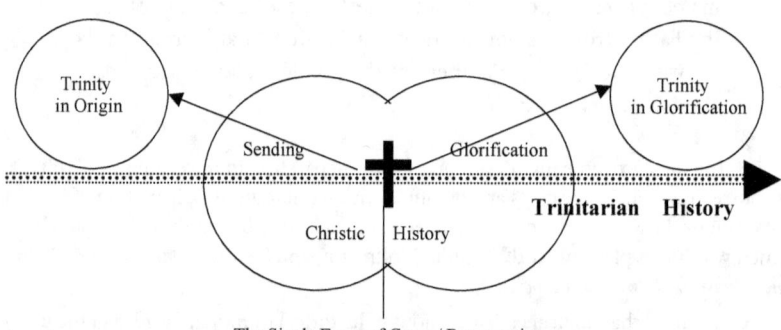

<Fig. 1: The conceptual relationship between Christic history and Trinitarian history>

---

51. *Trinity and the Kingdom*, 39.
52. Ibid.
53. *Church in the Power of the Spirit*, 62.

THE TRINITARIAN DOCTRINE OF CREATION

Panentheism in Trinitarian-Pneumatological Spatial Thinking

Doctrine of Creation Reconsidered in light of
Christian Trinitarianism

In *God in Creation*, Moltmann states that a Christian doctrine of creation views the world in light of Jesus Christ (the Messiah) and thus, views creation together with its messianic future, known since antiquity as "the kingdom of glory."[54] With this definition of the Christian doctrine of creation, Moltmann already demonstrates the direction his eschatological thoughts will take in his ecological doctrine of creation.

The symbol of "the kingdom of glory," a cosmic hope in nature, presupposes that "creation in the beginning" is an open creation, and that its consummation will be the dwelling place of God's glory. Therefore, in contrast with creationists' emphasis on the creation as what happened in the decisive past, Moltmann's concept of the Christian doctrine of creation demonstrates, from the beginning, a highly eschatological orientation.[55]

Another important basis of Moltmann's doctrine of creation is his social doctrine of the Trinity, characterized with a perichoretic communion and an equality of divine Persons. Moltmann reconsiders the creation in the perspective of this Trinitarian perspective and puts forth a "corresponding ecological doctrine of creation."[56] Instead of a monotheistic understanding of God as the one absolute subject, understanding God in a Trinitarian way—as the union of the Father, the Son, and the Spirit—broadens and enables a different view of creation. More concretely, this means two things. First, his Trinitarian perspective enables Moltmann to panentheistically see the divine immanence in the creation while maintaining God's transcendence. Second, the perichoretic inter-dwelling is eschatologically extended to the mutual influence and fellowship that is enabled through the Son by the Spirit in

---

54. *God in Creation*, 4–5.

55. "This messianic doctrine of creation therefore sees creation together with its future—the future for which it was made and in which it will be perfected." Ibid., 5; "If the cosmic Spirit is the Spirit of God, the universe cannot be viewed as a closed system. It has to be understood as a system that is open—open for God and for his future." Ibid., 103.

56. Ibid., 2.

the relationship between the Trinitarian God and the creation, as seen in Moltmann's view of eschatological Sabbath.[57]

With these qualifications of being Christian and being Trinitarian, Moltmann's revision of the traditional doctrine of creation proposes to emphasise the panentheistic relation between God and the creation together with their common future by virtue of God's self-giving, self-limitation, self-humiliation, and self-communication in the form of *zimsum* and *Shekinah*.

Zimsum: Divine Self-Contraction

With the help of Isaac Luria's idea of *zimsum*,[58] developed on the ancient Jewish doctrine of *Shekinah*,[59] Moltmann demonstrates a spatial thinking in his Trinitarian doctrine of creation. By applying the *Shekinah* idea to God and creation, Luria reasons that the existence of a world outside God becomes possible only by an inversion of God, a self-withdrawing of God's presence and power, which enables a "mystical primordial space," a theological tradition later called *nihil*, Nothingness, and God-forsaken space, hell, and absolute death.[60] Now that there came into being a space outside God, God can step out of Godself and can enter the newly created primordial space to creatively manifest God's loving nature. The spatial aspect of the idea of *zimsum* becomes more evident when Moltmann says:

> God does not create merely by calling something into existence, or by setting something afoot. In a more profound sense he "creates" by letting-be, by making room, and by withdrawing him-

---

57. Two meanings of the Greek term *perichoresis* have been transtated into two Latin terms: *circumincessio* (derived from *incedere*, meaning to permeate, com-penetrate and interpenetrate) and *circuminsessio* (derived from *sedere* and *sessio*, being seated, having its seat in, seat). Boff, *Trinity and Society*, 93, 134–36. In my opinion, *circuminsessio*, which I translate as inter-dwelling, is appropriate to describe the immanent Trinity and eschatological Sabbath because it can express the aesthetic joy in the fellowship. However, I would like to use *circumincessio*, which I translate as inter-permeation, is appropriate to describe the mission of Trinitarian God in history, because this term can emphasise the intentional penetration and permeation with its dynamic and historical flaovor.

58. "*Zimsum* means concentration and contraction, and signifies a withdrawing of oneself into oneself." *God in Creation*, 87.

59. This means God's indwelling in the creation, especially in the temple, by way of self-limiting and self-impartation.

60. *God in Creation*, 87.

self. The creative making is expressed in masculine metaphors. But the creative letting-be is better brought out through motherly categories."[61]

Here, God is no longer an unmoved mover but is seen to engage in a motion, contracting Godself into Godself to create a God-evacuated space for the creation to be in.

Creation in the Feminine Imagery of Conception and Birthing

Moltmann's concept of *zimsum* can best be described by the relationship between mother and child. The mother makes in herself room for the baby, wherein the baby could exist apart from the mother. The room is made possible by the mother who lets the baby exist; however, the space is still *in* her. Then, she enters into the space *within* her—a space she neither completely owns nor controls—for the benefit of the baby and reaches *out* with the umbilical cord through the space-within *to* the baby to give it life.[62] This image of the umbilical cord fits especially well with the idea that the creation is not a finished product, but rather an embryo, a fetus, or a baby, still being molded and in-formed and nurtured to grow inwardly and outwardly to reach its fullest expression. The Spirit as "the creative energy of God and the vital energy of everything that lives"[63] connects with the creation as a *life-line* to give it energy, form, and life. In this metaphor, Trinitarian panentheism acquires a new tangible pictorial sensibility.

What Moltmann hopes to achieve through this image developed from *zimsum* in a Trinitarian way is to offer a pictorial presentation of a doctrine of creation with an understanding of divine world-immanence while not discarding the traditional concept of God's transcendence. He claims that it is one-sided to hold only to the masculine creation metaphor of "the work of God's hands." Instead, he argues that "creation is also the differentiated presence of God the Spirit, the presence of the

---

61. Ibid., 88.

62. This becomes clearer when Moltmann describes the Holy Spirit in maternal terms: "The maternal mystery of the Holy Spirit contains the more intimate relationships of *outpouring, indwelling and mutual influence*. 'Grace and the gifts of grace', the 'Spirit and the powers of the Spirit', form community and *permeate creatures without alienating them or destroying them*." Moltmann-Wendel and Moltmann, *God—His and Hers*, 37. Emphases are mine.

63. *Way of Jesus Christ*, 91.

One *in* the many."⁶⁴ This feminine pneumatological remark on the mode of divine immanence in creation presupposes and points to the juxtaposed and double-layered divine relationship of transcendence and immanence. Moltmann lovingly offers both a respectful separation of God from the creation to give it existence and freedom alongside a nurturing and life-giving connection, so that it won't perish into non-being again.⁶⁵

In this female-biological picture, we also begin to understand Moltmann's spatial riddle of outside and inside concerning the relationship of God and the creation: the creation is outside God and at the same time it is in the Trinitarian God. In Moltmann's words, "If creation *ad extra* takes place in the space freed by God himself, then in this case the reality outside God still remains in the God who has yielded up that 'outwards' *in* himself."⁶⁶ This spatial riddle becomes intelligible when we think in terms of motherly *begetting* of the creation.

First, in this female-biological picture, *zimsum* as God's creative suffering becomes more analogically understandable. Making a space in oneself that does not totally belong to oneself for someone else both presupposes and entails physical and psychological burdening in the mother. The letting-be of this space, as well as that of the thing occupying this space, are both the result of, and incurs more of, self-limitation and self-sacrifice on the part of the mother or the Creator.

Second, in this female-biological picture, the continuity and discontinuity in the relationship of the Father to the Son and to the creation become intelligible. As the Son was begotten eternally, the creation will be begotten in time in God's womb.⁶⁷ Moltmann hinted at the continuity and discontinuity between the Son and the creation. On the one hand, Moltmann tries to understand the begetting of the Son as a birth from

---

64. *God in Creation*, 14. Emphasis is the author's.

65. Ibid. Here Moltmann contrasts two modes of God's relationship to the world: one of transcendence (making, preserving, maintaining, and perfecting) and one of immanence (indwelling, sympathising, participating, accompanying, enduring, delighting, and glorifying). The latter Moltmann calls relationships of mutuality that may surely suggest a motherly nurturing of the baby inside.

66. *God in Creation*, 88–89. Emphases are the author's.

67. This thinking corresponds to an intimate relationship between the immanent Trinity and the economic Trinity in Moltmann, explained in this chapter in the section titled "Inseparable Mutual Relationship Between Immanent and Economic Trinity" in this chapter.

His motherly Father.⁶⁸ On the other hand, although Moltmann never forgets the eternal begetting and thus the uniqueness of the Son, nevertheless the begetting of the creation is aligned along the same loving outreach of the Father toward the Son:

> The creation comes into being in the same line of fatherly / motherly love toward the Son. The love of the Father which begets and brings forth the Son is therefore open for further response through creations which correspond to the Son, which enter into harmony with his responsive love and thereby fulfil the joy of the Father. Hence the love of the Father which brings forth the Son in eternity becomes *creative* love. . . . Creation proceeds from the Father's love for the eternal Son.⁶⁹

Therefore, the creation comes into being out of the self-giving, self-communicating, going-out-of-Godself love of the Father who, beyond the begetting of the Same, creates something other than Godself (so, not the Son). In return, the creation responds to the Father's initiating love *with* the Son and *in* the Son. In the sacrificial godless space for "the other," the creation was called into existence in the life-giving and sustaining motherly embrace of the Holy Spirit. This analogous continuity forms the logical basis of 1) the incarnation of the Son and also 2) the regeneration of the believers in the Spirit in the same manner with the pneumatological event of "rebirth to life" of the Son at Easter.⁷⁰

Third, in this female-biological picture, the transience and death that is inherent in nature and became eschatologically conquered in Christ's resurrection becomes more intelligible. For Moltmann, nature is "the reality of that world which is no longer God's good creation and not yet God's kingdom," ⁷¹ perceived as a "victim to transience and death"⁷² and "the enslaved creation that hopes for liberty."⁷³ Moltmann, however,

---

68. "If the Son proceeded from the Father alone, then this has to be conceived of both as a begetting and as a birth. And this means a radical transformation of the Father image; a father who both begets and bears his son is not merely a father in the male sense. He is a motherly father too. . . . He has to be understood as the motherly Father of the only Son he has brought forth, and at the same time as the fatherly Mother of his only begotten Son." *Trinity and the Kingdom*, 164.

69. Ibid., 168. Emphasis is mine.

70. *Spirit of Life*, 144–60; *Way of Jesus Christ*, 247–50.

71. *God in Creation*, 38.

72. *God in Creation*, 68; *Way of Jesus Christ*, 169–70.

73. Ibid., 21; *Crucified God*, 101.

would consider it highly dubious and an anthropocentric idea that this transience and death intrinsic to nature is due to human sin; rather, for Moltmann, nature is subject to transience and death because it is *"alienated from the source of its life* and *in the imprisonment of the Spirit* that animates it."[74] In this connection, one of Moltmann's favourite texts comes forward: "When God withdraws his Spirit, it is the same as if he turns away, and hides his face. It means death for the human being and for everything else that God has created (Ps. 104)."[75] The "petrified" and "frozen" condition of the present time nature is the result of the harmful and threatening environment of *nihil* and the separation from the divine energy that brings life to all living things.[76]

This view on nature as intrinsically bearing death and transience is perhaps one of the points where Moltmann is most severely criticized. Douglas J. Schuurman, for example, accuses Moltmann of failing to distinguish creation and fall and, therefore inevitably arrives at a creation-annihilating vision of Eschaton.[77] Similarly, Steven Bouma-Prediger argues that Moltmann views creation as necessarily faulted and thus a new creation is essential.[78] Moltmann is even accused of being Gnostic

---

74. *Way of Jesus Christ*, 253, emphases are mine; "Having called creation in the beginning a system open for time and potentiality, we can understand sin and slavery as the self-closing of open systems against their own time and their own potentialities." *Future of Creation*, 122.

75. *Spirit of Life*, 41, 45, 288.

76. The terms "petrified" and "frozen" that Moltmann uses to describe nature should not be taken as his general description of nature. When he describes nature with these terms, he refers to the cruel and dark side of the present state of the creation, which he interprets as not yet fully Spirit-filled. To use the metaphor of the creation as a baby again, the creation is a beautiful baby and yet it has to keep growing to reach its full potential, depending on the continuous in-breathing of the Spirit, the source of all beings. Moltmann tends to emphasise this "not yet" side of nature in an eschatological perspective *vis-a-vis* the creationists' over-emphasis on the already perfected creation. On the other hand, there are abundant signs of the life-giving Spirit in the creation and I do not think that Moltmann would argue against it. So, he describes the two sides of the creation in relation to the presence of the Spirit, as he quotes Psalm 104. The Psalm itself is full of wonder and praise of the abundance and majesty of God's creation although it contains the frozen and petrified side of creation, as in verse 29 ("When you hide your face, they are terrified; when you take away their breath, they die and return to the dust"). Moltmann justifiably interprets this as creation's "not yet" side and its need of and dependence on the Spirit of life.

77. Schuurman, "Creation, Eschaton, and Ethics," 67.

78. Bouma-Prediger, "Creation as the Home of God," 89.

since he assumes the world *qua* world requires redemption.[79] To a certain degree, these criticisms are valid because Moltmann does conflate creation and fall, and in a sense God seems to have no other choice but to provide, on top of the forgiveness of human sin, a new creation as an adequate solution to this "faulted" creation that causes universal suffering and sin.

Moltmann's choice, however, does have a theological strength of its own and needs to be defended. First, interpreting the Fall as an historical incident and insisting on the existence of a pristine state of creation at the Garden tends to be a naïvely literal interpretation of the Scripture, both in light of interpretation history of Genesis 3 and modern science.[80] In addition, science tells us that there surely were sickness, death, and numerous extinctions caused by disease, earthquakes, and volcanoes long before humans stood on earth and sinned. Second, viewing the creation as incomplete and open to, and longing for, the future glorification is not Gnostic but a basic biblical perspective on nature (Romans 8). Furthermore, this perspective means neither negating the goodness of creation nor viewing it as faulted. Creaturely finitude, as in a newborn baby, can be both good and incomplete, and open to eschatological consummation. Third, Moltmann's account of the relationship between sin and death that the transience and finitude of creation entails, provides a more realistic understanding of suffering than the simple doctrinal repetition that all creaturely suffering is caused by human sin.[81]

Having said that, Moltmann's hamartiology does need greater development. The most striking aspect about Moltmann's hamartiology, I suggest, is how little he talks about sin—compared to how much he talks about suffering—rather than his definition of sin in relation to eschatological goal and the Holy Spirit.[82] Perhaps Moltmann's intention here is not to justify suffering as a legitimate punishment of sin. As a life-long political theologian, he is well aware that there is a mountain of difference between the suffering of the victims and suffering experienced by

---

79. Farrow, "Review Essay: In the End is the Beginning," 438.
80. Wiley, *Original Sin*.
81. *Spirit of Life*, 125ff. *Coming of God*, 77ff., 306f.
82. In fact, Moltmann's definition of sin as self-closedness of creaturely beings and systems, an alienation from the source of its life does not sound strikingly different from another Reformed theologian Karl Barth's definition of sin as the rejection of grace. Mangina, *Karl Barth*, 127.

perpetrators; as well, for so long the church has only been interested in the absolution of sin of the oppressors without heeding to the pain of the violated and oppressed.[83] The predominance of the discourse on sin in the face of planetary suffering blurs the fact that there are different degrees of accountability for the suffering. Suffering can be caused by humans—like Auschwitz and Hiroshima—and nature—like the painful and fearful experience of death by every living being on earth by diseases and disasters. In the face of such different kinds and degrees of suffering, a universal and quick diagnosis would suffice neither as consolation for the victims nor a solution for the suffering.[84] Instead of such an indifferent and insensitive hamartiology as "we all sinned and we all deserve suffering," we need a more differentiated and sensitive discourse on suffering and the doctrine of sin that can suggest different routes to liberation for the oppressed and the oppressor. In this connection, Moltmann seems to have opted for a more compassionate discourse toward the victims of suffering 1) by not universalizing sin and thus justifying suffering as the legitimate punishment and 2) by proposing God as demonstrating divine compassion and solidarity on the cross.

Moltmann's choice, however, does not negate the need for a more detailed discourse on sin, because universalizing suffering creates as many problems as universalizing sin. By refusing to differentiate between the suffering as a result of sin and the suffering as a result of *nihil*, even though the latter is a more fundamental dimension, Moltmann fumbles the opportunity to address human accountability regarding human-caused suffering, such as social-structural sin, genocide and ecological destruction experienced by the poorest of the world. I also would like to point out that the victims of suffering are not and do not remain entirely innocent from sin—like the impoverished Indonesian peasants who as a last resort chose to burn forests to grow crops. In the face of such a convoluted relation between suffering and sin, we need a more sophisticated and targeted discourse. How can a discourse on sin become a more targeted discourse on ecology, and more importantly, how can it lead people to repent? For that purpose, I believe that Moltmann needs to bring his eschatological / ecological definition of sin as self-isolation in line with his political theology.

---

83. *Spirit of Life*, 126.

84. Funeral sermons that refer to sin as the cause of death are really the advice of Job's friends, even though we are all sinners.

In sum, the creation's location in, and relationship to, the *nihil* and the Trinitarian God can be expressed in the following way: the ontological bond of the Holy Spirit holding (and giving life to) the creation, while not completely protecting it from the chaotic and nihilistic power, is expressed by the bridge-like umbilical lines.

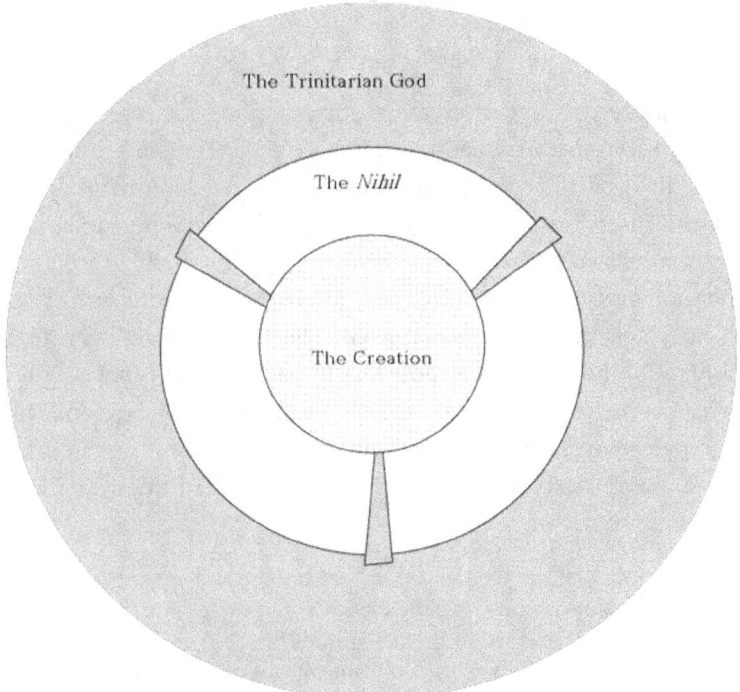

<Fig. 2: *Zimsum* and the umbilical connection of the Spirit>

The Creation Open for the Kingdom of Glory

Three Kinds of Creation

For Moltmann, creation is neither a one-time act of God a long time ago nor an already perfected complex system with a perfectly harmonious order. In fact, Moltmann distinguishes three types of creation: 1) *creatio originalis,* or God's initial creation; 2) *creatio continua,* or creation in history or God's historical creation; and 3) *creatio* nova, or the creation of the End-time or God's "perfected creation."[85] Moltmann elaborates:

> We can see initial creation as the divine creation that is without any prior conditions: *creatio ex nihilo*; while creation in history is the laborious creation of salvation out of the overcoming of disaster. The eschatological creation of the kingdom of glory, finally, proceeds from the vanquishing of sin and death, that is to say, the annihilating Nothingness.[86]

This tripartite concept of creation corresponds to the eschatological thinking of Moltmann already seen in his doctrine of the Trinity.[87]

In his tripartite concept of creation, however, Moltmann does not grant the three parts equal theological weight. Although each of the three phases of creation is important and made possible through the Trinitarian participation of God, creation in the beginning and creation in history are fundamentally oriented toward the future glory. In this regard, it is important here to note that the future glory is not simply the recovery of the initial creation; as well, the continuous creation is more than repairing work, because the initial creation itself is not a perfectly ordered closed system, but an open system oriented toward its future. Moltmann notes;

> The notion of a perfect, self-sufficient equilibrium in the resting, stable cosmos contradicts the biblical—and even more the messianic—view of creation aligned towards future glory. The idea of the future as a *restitutio in integrum* and a return to the original paradisal condition of creation (*statis integritatis*) can neither be called biblical nor Christian.[88]

Therefore, it is the new creation, or the eschatological goal of the kingdom of glory, that gives the whole process, including the other two creations, meaning and direction. For God will be present in the cre-

---

85. *God in Creation*, 55; *Way of Jesus Christ*, 286–87. This expression of perfected creation could be used to enhance the sense of continuity among the three kinds of creation in line with the analogy of infant creation coming forth from the divine womb and maturation and consummation of that creation.

86. *God in Creation*, 90.

87. Ibid., 2.

88. Ibid., 208.

ation in a new, direct and unmediated way,[89] by "the indwelling of God in this new world," making the creation "the home of the Trinity."[90]

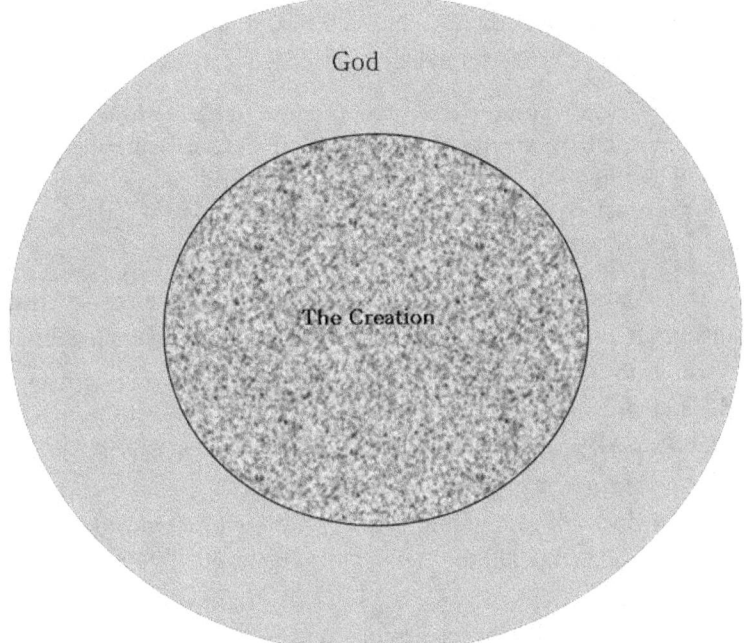

<Fig. 3. The creation wholly embraced and infiltrated by God>

Eschatological Realignment of Nature and Grace

Moltmann maintains that traditional theology has emphasized duality between creation and redemption, creation and covenant, nature and super-nature, necessity and freedom. This two-tiered thinking was formulated in the famous medieval principle: *gratia non destruit, sed praesupponit et perfecit naturam* (grace does not destroy but builds upon and perfects nature).[91] Moltmann points out that this maxim starts from the assumption that "the grace of God is to be seen in the incarnation of the eternal Logos in Christ, and concludes that this incarnation presupposes and perfects creation."[92] In line with this thinking, Rahner could say

---

89. Ibid., 287.
90. *Trinity and the Kingdom*, 104.
91. *God in Creation*, 7.
92. Ibid.

that "anthropology is 'deficient Christology' and Christology is 'realised anthropology.'"[93]

Moltmann, however, points to the failure of this maxim to distinguish between grace and glory, between history and new creation, and between being a Christian and being perfected:

> [T]he glory which perfects nature is supposed already to be inherent in the grace; the kingdom which is the inner foundation of creation is thought already to exist in the covenant; and the perfection of the human condition is considered to be already integral to being a Christian.[94]

This failure, which continually leads to arrogant ecclesial triumphalism, surely has everything to do with the lack of eschatological perspective. Moltmann proposes to reformulate the second part of the medieval theological tenet in the eschatological outlook of the creation: *Gratia non perfecit, sed praeparat naturam ad gloriam aeternam; Gratia non est perfectio naturae, sed praeparatio messianica mundi ad regnum Dei* (Grace does not perfect, but prepares nature for eternal glory; grace is not perfected nature but the messianic preparation of the world for the Reign of God).[95]

In this eschatological reformulation, both nature and grace are made relative in light of something yet to come. Because Christ is the beginning of the new creation of the world, "We have to talk about nature and grace, and the relationship between nature and grace, in a forward perspective, in light of the coming glory, which will complete both nature and grace, and hence already determines the relationship between the two here and now."[96] Therefore, Moltmann states that nature and grace complement each other *in preparation for* the coming glory, instead of a hierarchical, dichotomous structure of grace over nature.

The significance of overcoming this dualistic structure of nature and grace in an eschatological relativisation of both nature and grace, Moltmann believes, is that it is possible "to discern and define more precisely the possible reconciliation between freedom and necessity, grace and nature, covenant and creation, being a Christian and being a human

---

93. Ibid.
94. Ibid.
95. Ibid., 8.
96. Ibid.

being."⁹⁷ In contrast, traditional dualism not only instills triumphalism in Christians but also fosters an indifference and apathy toward the natural world. Placing the church in a larger framework—the threefold creation, as opposed to the oppositional, dualistic structure of nature and grace—Moltmann calls the church to a newly recognized eschatological mission that looks forward to and anticipates the new creation.

The World as Promise and Anticipation

Barth developed the idea of the world as a parable of the Reign of Heaven, in which the everyday experiences of the world become pointers to something different. For this idea, he uses a theatrical metaphor, in which the world becomes the theatre and setting, the location and background, with the characteristics of constancy, of rotation, and of persistence.[98]

Moltmann evaluates Barth's theory as an attempt to bridge the undue duality of covenant and creation, but at the same time he criticises Barth for belittling the creation as "'only' the theatre, 'only' the space in which God's own glory shines forth in the work of reconciliation."[99] Moltmann thinks Barth's failure is two-fold: first, Barth thought of only creation in the beginning and its preservation, not the continuing and contingent history of creation; and second, Barth extols "the revelation of reconciliation" as being already "the triumph of . . . glory."[100] In short, Moltmann thought Barth was not eschatological enough.

Instead of Barth's idea of the world as a parable of the Reign of Heaven, Moltmann proposes an understanding of the world as an historical parable of its own future.[101] He proceeds to say that the creation is more than a mere parable—the creation is the promise and anticipation of what is to come, in that it points towards its own fulfilment and anticipates a future still to come.[102] Thus Moltmann could declare: "If the world as creation is the real promise of the kingdom of God, it then itself belongs to the history of the kingdom and is not merely its 'stage and backcloth.'"[103] As an historical parable, that is, a promise and anticipation

97. Ibid., 9.
98. Ibid., 60–61.
99. Ibid., 61.
100. Ibid., 62.
101. Ibid., 61–62.
102. Ibid., 63.
103. Ibid., 63.

of the coming Reign of God, the creation is open to, and indeed is in the process of becoming, the new creation.

Creator Spiritus

If we use a Trinitarian-pneumatological paradigm for creation, and consider it as a divine birthing instead of the masculine product of a machine, and if it is true that the creation is not a ready-made closed system but an open system directed toward future glory, then we need to articulate the mode of divine presence in the creation. Does the divine presence cancel the difference between God and the creation and justify whatever already exists? Or is the presence of the Spirit a proof of the creation's imperfection and a protest against it? What is the role of the Spirit in the creation? Does the presence of the Spirit preserve and sustain the creation or does it transform the creation and guide it into its future destination? The previous two sections already shed light on these questions. But at the same time these themes also need to be clarified in light of the mode of presence and the role of the Spirit in creation. In this section, I will articulate the mode of the divine immanence of the Holy Spirit in creation in the context of Moltmann's Trinitarian panentheism. I will show that this particular understanding of the Holy Spirit helps us to comprehend the Holy Spirit's seemingly contradictory works.

The Spirit as the Ontological Bond in Trinitarian Panentheism

The doctrine of *zimsum* encourages us to think of the creation in God without necessarily falling victims to pantheism: If the creation is born in the primordial space of *nihil*—a God-evacuated, independent, space still present in God[104]—how can Trinitarian panentheism articulate God's relation to the world? This question belongs both to Trinitarian pneumatology and Christology and, in my opinion, Moltmann's theological genius is most evident here.

The creation that is created in the Godforsaken space of *nihil* is not yet redeemed from the threat of returning to nothingness, but is still in God. The inner secret of the contingent creation that does not have the intrinsic right and power to exist, and yet still mysteriously exists, is the Spirit of God. I already explained this mystery using terms of the feminine imagery of conception. The creation in *nihil* is con-

---

104. Moltmann describes the *zimsum*-based creation as "the creation *out of* God *in* God" (aus Gott in Gott). *Trinity and the Kingdom*, 111.

nected to God through the Spirit, who holds and sustains the creation, protecting it from the threat of *nihil*, giving it existence and life. The Spirit is the umbilical cord and ontological bond between the creation and the Trinitarian God. The Spirit touches, comforts, cuddles, and nurtures the creation, filling it with divine immanence and at the same time continuously moulds, transforms, awakens and guides it toward the transcendent intention and goal that can only be found in the loving fellowship of God with the creation in the kingdom of glory. The Trinitarian-pneumatological concept of divine presence binds together God's transcendence and immanence.

This motherly caring for, and loving embrace of the creation, is how the Spirit of God is present in creation. This presence does not violate the freedom of the creation by intruding and imposing alien order, and yet it does not desert the creation, protecting it from threats and helping it realize its future potential. Therefore, the creation is in the God-less environment, in a harsh reality of separation from, or lack of the immediate presence of God; but on the other hand, it is in the presence of the Spirit and thus in God, waiting for its future in which the creation reaches the consummation and the distance between it and its Creator is abolished.

"Already" and "Not Yet" of the Spirit

This mode of God's presence, as well as the creation's dual location in terms of its environment (in *nihil* and in God) leads us to the question of "already" and "not yet" of the Spirit. Is the Spirit already in the creation and thus can be identified with the cosmic Spirit? Or does the Spirit exist only as a promise and an agent for the transformation of the world?

In times of incredibly beautiful natural phenomena, such as the blossoming of trees and flowers in the Spring, the gentle breeze along the beach against the background of a sunset during the Summer; the blazing autumn colours against an unbelievably blue sky; and the ineffable Canadian snow that fills the whole wide empty sky, we feel the creation is filled with the Spirit. But there are other moments when nature seems so strange, empty, cruel, and indifferent to our human feelings and aspirations. Earthquakes, hurricanes, volcanic eruptions and tsunamis too often entail massive and indiscriminating death and grief; the wolf must eat the lamb to live another day; the "survival of the fittest" reveals nature to be, in the words of Alfred Lord Tennyson, "red in tooth and claw."¹⁰⁵ How can we Christians speak of the presence of God in these phenomena, especially in light of Christ, who died as a victim of violence and a victim of the death of all living things?

Denis Edwards identifies two approaches in this matter: a "sacramental" theology of divine presence and a "prophetic-eschatological" approach.¹⁰⁶ Theologians belonging to the former approach, such as Yves Congar, John V. Taylor, Elizabeth Johnson, Wolfhart Pannenberg, see the Holy Spirit everywhere working in every possible realm and level of the creation; whereas theologians adhering to the latter, such as Michael Welker, John Zizioulas emphasize the "not yet" of the Spirit, eschatologically reserving the presence of the Spirit in the creation.

Moltmann, however, according to Denis Edwards subscribes to both a "sacramental" theology of divine presence and a "prophetic-eschatological" approach in the relationship of the Spirit to creation.¹⁰⁷ Affirming the presence of the Spirit in creation Moltmann says: "The possibility of perceiving God in all things, and all things in God, is grounded theologically on an understanding of the Spirit of God as the power of creation and the wellspring of life."¹⁰⁸ At the same time, he emphasizes the unredeemed state of the creation so expressly that he talks about "creation as a community of suffering," and "a kind of winter of creation."¹⁰⁹ He looks to an eschatological coming of the Spirit who will transform, heal, redeem, and overturn the transience, death, and destructiveness of the creation.

How does Moltmann synthesize these two aspects of the presence of the Spirit in the creation? What is the relationship of the cosmic Spirit and the eschatological Holy Spirit? How can we conceive the mode of presence and work of the Spirit of God in the scheme of Trinitarian-pneumatological panentheism?

The Eschatological Spirit and the Cosmic Spirit

Moltmann distinguishes four different types of efficacy of the Spirit: creating, preserving, renewing, and consummating.¹¹⁰ These four types

---

105. Ross, *Alfred, Lord Tennyson In Memoriam*, 36.
106. Edwards, "Ecology and the Holy Spirit," 42–159.
107. Ibid., 143.
108. Moltmann, *Source of Life:*, 120; *Spirit of Life*, 35.
109. Ibid., 122.
110. *God in Creation*, 12.

can be regrouped into two: the works of the immanent cosmic Spirit (creating and preserving activities), and the works of the eschatological Spirit (renewing and consummating activities). Thus Denis Edwards is right when he argues that Moltmann upholds both a "sacramental" theology of divine presence and a "prophetic-eschatological" approach. Moltmann differentiates and affirms these two aspects of the work of the Spirit. But because the creation was from the beginning open to its future glory, the focus of the work of the Spirit is the eschatological transformation, although the work of initial creation and preservation is always presupposed or hind-sighted. In Moltmann's words:

> Theologically [the Spirit of creation, preservation and development] must be called the Spirit of God and the presence of God in the creature he has made. But according to biblical usage, this is not the Holy Spirit. The Holy Spirit is the name given to the Spirit of redemption and sanctification. . . . "The Holy Spirit" does not supersede the Spirit of creation but transforms it.[111]

Both of the two kinds of work of the Spirit are experienced in the *creatio continua*. It means "God's activity in preserving creation from the powers of annihilation" and "the continuous sustaining of the creation which was once brought into being."[112] But, in prophetic theology, we can find a second meaning of creation in history: "God's historical activity is directed, not towards the preservation of what was once created, but towards the anticipation of the salvation in which creation will be consummated."[113] Continuous creation in this second sense is not only *creatio continua* but simultaneously *creatio nova* and *creatio anticipativa*.[114]

How can we understand Moltmann's juxtaposition of the Holy Spirit's two areas of work, which some theologians deem inappropriate?[115]

---

111. Ibid., 263.
112. Ibid., 208–9.
113. Ibid., 209.
114. Ibid.
115. Farrow deems Moltmann's special concern to hold together immanence and transcendence of the Spirit as a rehabilitating emanationism of the Christian neoplatonists mainly on the ground that the attribution of preserving creation to the Spirit, using the terms "energy" and "efficaciousness", undermines the Personhood of the Spirit. Farrow, "Review Essay: In the End is the Beginning," 431. Deane-Drummond also criticizes Moltmann for not being able to adequately affirm the immanence of the Spirit, by failing to make an appropriate distinction between the work of the Spirit in creation

How can we understand the seemingly self-contradicting dual agency of the Holy Spirit, that is, the preservation and transformation? How can we combine without contradiction the prophetic-eschatological understanding of the Holy Spirit as the protesting and changing agency with the understanding of the Holy Spirit as the world-sustaining immanent God? These questions are important in order to affirm the intrinsic value of the creation apart from a utilitarian perspective and at the same time, resist an oppressive political conservatism that legitimizes whatever already exists.

I suggest once again we employ the metaphor of a child *in utero* to understand the mode of presence and the work of the Spirit in the creation. Whether Moltmann keeps this image throughout his thinking on the doctrine of creation or not, the womb helps us understand not only Moltmann's spatial thinking, but also the dual agency of the Holy Spirit.

Moltmann's basic claim in his pneumatological doctrine of creation is that through the Holy Spirit, God is present in the creation. The point of *ecological* thinking about the creation is to see the world in the environment of the Trinitarian God. By envisioning the Holy Spirit here as the ontological bond and umbilical cord between God and the creation—in the God-evacuated primordial space of *nihil*—the mode of presence and dual efficacy of the Holy Spirit can be more biologically understood. God is not the creation and the creation is not God; but God is present in an essential way, giving it existence, freedom, and the power to grow—in a word, life.

But if we envision the creation as an open system, that needs to travel into its future, that needs to be nurtured to have the strength to do so, in a relatively autonomous space in its own order with its own pace, the Holy Spirit cannot be an alien intruder of the creation's inner logic and imposer of a ready-made heavenly order. The presence of the Holy Spirit, like the presence of the mother to an embryonic baby, must be a nurturing and guiding presence, without ceasing to hold and sustain

---

and in redemtpion. She thinks that this pneumatological synthesis of Moltmann is, while inspiring, a main hindrance from making a decisive emphasis on the work of the cosmic elements in the work of the Holy Spirit, leaving scattered remarks in *God in Creation* or *Coming of God*. Deane-Drummond, *Creation through Wisdom*, 118–20, 138. I disagree with both authors. They fail to discern Moltmann's demonstration of immanent transcendence in the pictorial image of bearing and birthing of a child, which is neither emanationism, nor does it fail to adequately emphasise the material/physical dimension of the Spirit's work.

patiently in whatever state the baby has arrived. But the nurturing and guiding presence is, in a fundamental sense, an eschatological presence; if the power and wisdom to arrive at its future does not lie intrinsically in the creation but can be given through the interaction (theologically identified as inspiration) with the Spirit-parent, then the dual agency of the Spirit in sustaining and transforming can be quite understandable.

THE GLORIFICATION OF GOD IN THE COSMOS AS THE CLIMAX OF TRINITARIAN HISTORY

Unifying Principle of Perichoresis *in Moltmann's Theology*

Perichoresis in Trinity

Probably one of the most salient characteristics of Moltmann's theology is his doctrine of God as the social Trinity. Unlike those who take the traditional Western approach to the Trinity, he starts with the threeness of God instead of one dominant Person as the source and origin of all divinity, thus safeguarding the distinctiveness of the Persons. As a result of such a starting point, the unity, the oneness of the Persons might be at stake and becomes crucial to his theology.

For Moltmann, the unity of the Three comes from the interpenetrating, interdwelling communion of the three Persons, called *perichoresis*.[116] The Three live in community because of this kind of harmonious dancing, that is, they live in love dynamically overflowing among them. By using *perichoresis*, Moltmann avoids tritheism, while nevertheless starting with distinctive three Persons. Thus, *perichoresis* is a key term in Moltmann's doctrine of the Trinity.

The importance of *perichoresis* is that it enables the unity of the Three without losing its diversity; it enables the oneness without amalgamating the three or losing individual characteristics.[117] On the other hand, it establishes the identity of each in relationship to the others. In this way, *perichoresis* enables co-establishment of identity and relationship. This aspect of *perichoresis* is important for other concepts: human

---

116. *Trinity and the Kingdom*, 174–76.

117. "The doctrine of the *perichoresis* links together in a brilliant way the threeness and the unity, without reducing the threeness to the unity, or dissolving the unity of the threeness." Ibid., 175.

beings as the *perichoretic* unity of body and spirit; the creation-community, consisting of human beings and the other creatures in their multi-strata composition and communication; and finally the eschatological panentheistic Sabbath. Moltmann expresses it this way:

> Our starting point here is that all relationships which are analogous to God reflect the primal, reciprocal indwelling and mutual interpenetration of the Trinitarian *perichoresis*: God in the world and the world in God; heaven and earth in the kingdom of God, pervaded by his glory; soul and body united in the life-giving Spirit to a human whole; woman and man in the kingdom of unconditional and unconditioned love, freed to be true and complete human beings.[118]

In this way, Moltmann expands the Trinitarian concept of *perichoresis* to describe the theological significance of an eschatological communion between God and the creation and among the creation community.

Perichoresis in the Doctrine of Creation

According to Moltmann, the biblical traditions tell us that the earth and the bodily creatures are the end result of God's work in creation.[119] He writes, "The earth is the object and the scene of the Creator's fertile and inventive love,"[120] which is the *perichoretic* union of Trinitarian Persons. In that sense, for Moltmann, creation is the outward expression and the end result of divine *perichoresis* i.e. the immanent Trinity. The inner-Trinitarian *perichoresis* is the pattern and model in creating the world and in God's relationship with the creation. Moltmann says:

> This means that we have not, either, understood the relationship of the triune God to the creation of his love as a one-sided relationship of domination. In contemplating the wealth of this eternal love, we have seen it as a complex relationship of fellowship; and in this complexity and plurality as therefore also reciprocal. 'Creation in the Spirit of God' is an understanding which does not merely set creation over against God. It also simultaneously takes creation into God, though without divin-

---

118. *God in Creation*, 17.
119. Ibid., 245.
120. Ibid.

ising it. In the creative and life-giving powers of the Spirit, God *pervades* his creation.[121]

In short, the relation of God *to* the creation is not one-sided, but a reciprocal, *perichoretic* relationship.

But Moltmann does not maintain that the creation is currently in a *perichoretic* communion with God; rather, in his overall theology and especially in *The Crucified God*, he is utterly mindful of the fact that this world is still unredeemed, and in the midst of suffering, alienation and death. But, for God's part, it can be said that God approaches the creation with the intention and expectation of *perichoretic* love. Of course the creation is not completely there yet.

When it comes to the *eschaton*, however, Moltmann envisions a state in which the world will be taken into a complete *perichoretic* union with God. The creative Spirit assumes the creation into God so that God fulfills God's creation, becoming all in all, making the creation pan-en-theistic. Moltmann boldly states: "In his Sabbath rest he allows his creatures to exert an influence on him. From the aspect of the Spirit in creation, the relationship of God and the world must also be viewed as a *perichoretic* relationship."[122]

Perichoresis in Anthropology

Moltmann's principle of *perichoresis* also plays a crucial role in his anthropology. Moltmann thinks that behind the traditional Christian overvaluation of man vs. woman, spirit vs. body and heaven vs. earth, there lies a serious Gnostic dichotomy between body and soul. Moltmann traces the origin of this separation to the psychological reduction of *imago Trinitatis* by Augustine and Thomas Aquinas, in which "the human being corresponds to the single Being of the triune God, not to the threefold nature of God's inner essence."[123] This reduction regards the spirit as man's essence and his true value, while the body is considered as a submissive and insignificant material, or, sometimes, even a prison from which to escape. In this way, the reductionist conception of *imago Dei* inevitably establishes an anthropology in which the soul has primacy over the body. Moltmann sees this kind of alienation from

---

121. Ibid., 258. Emphasis is the author's.
122. Ibid.
123. Ibid., 235.

the body as "the inner aspect of the external ecological crisis of modern industrial society."[124]

For Moltmann, this is an illegitimate separation; he maintains that body and soul form an inseparable unity through "a *perichoretic* relationship of mutual interpenetration and differentiated unity,"[125] rejecting both the primacy of the soul and that of the body. The body is no longer a subservient, despicable piece of flesh but an indispensable, important partner with the spirit. This presentation of the relationship between body and soul echoes Moltmann's explanation of the Three Persons in the Trinity.

Furthermore, Moltmann suggests that this concept of community should be the archetype of true human community.[126] Of course this expansion of a *perichoretic* community will embrace nature to constitute the creation community, which again will, as the home and indwelling place of God, be in *perichoretic* union with God in the eschatological panentheistic consummation, called Sabbath.

Eschatological Perichoretic Union of God with the Cosmos

Embodiment as the Goal of God's Work

Moltmann argues that, according to the biblical traditions, embodiment is the end of God's work in creation, reconciliation, and redemption.[127] The creation process moved from the resolve to the word, from the word to the act, and from the act to created reality. The world was reconciled in the movement in which the Word became flesh; the redemption of the world will be completed in "the new earth" and the "transfigured" embodiment is the fulfillment of the yearning of the Spirit (Romans 8).[128]

Note, however, that these divine movements toward embodiment are not intended to create an object separated from and over against the Creator, unlike the way a human creation of a desk or a watch could be alienated from their human creator. Embodiment moves toward a oneness, a harmony, and a peace, a perichoretic union between the Creator / Reconciler / Redeemer and the creation. Therefore, the eschatological

124. Ibid., 48.
125. Ibid., 259.
126. Ibid., 234–43.
127. Ibid., 244–46.
128. Ibid., 245–46.

procession of both God and the creation is a procession toward embodiment in which God and the creation arrive at a peaceful, harmonious, and loving union, in fact, a *perichoretic* union, which Moltmann calls God's rest.

The First Sabbath of God

According to Moltmann, biblical traditions show that on the Sabbath and through the Sabbath, God completed God's creation and blessed it.[129] Although the Western tradition has overlooked the importance of this seventh day, biblical traditions demonstrate that "the whole work of creation during the six days was performed *for the sake of* the Sabbath."[130] Another important point Moltmann notes in the biblical traditions is that although each day was followed by a night, God's Sabbath knows no night but becomes the feast without end.[131] These points already indicate the eschatological significance of Sabbath.

Moltmann describes the first Sabbath of God in three aspects. First, this rest is "*from* all his work which he had done" (Gen 2:3).[132] In this Sabbath, God returns from the work of creation to God's self, but not "to the world-less, eternal glory which precedes creation"; God returns to God's self along with the creation. So God's rest simultaneously becomes the rest of God's creation; and God's good pleasure in the creation becomes the joy of created things themselves.[133]

Second, God's rest is a rest *before the face* of God's works.[134] In God's rest, God lets the creation exist before God's face and co-exist with God's self so that the creation can acquire its real existence. By God's rest, God allows the beings God has created, each in its own way, to act on God. In other words, through God's rest, God allows God's self to be affected, to be touched by each of God's creatures. Therefore, with God's Sabbath of creation, God's history with the world begins, as does the world's history with God.[135]

---

129. Ibid., 276.
130. Ibid., 277. The emphasis is mine.
131. Ibid., 277.
132. Ibid., 278. The emphasis is the author's.
133. Ibid., 278–79.
134. Ibid., 279. The emphasis is the author's.
135. Ibid., 279.

Third, God also rests *in* the creation.[136] The resting God is wholly present in the world. Although the works of creation display the Creator's continual transcendence over the creation, the Sabbath of creation points to the Creator's *immanence* in creation. In the Sabbath "God joins his eternal presence to his temporal creation and, by virtue of his rest, is there, with that creation and in it."[137]

The Eschatological Significance of God's Sabbath

These aspects of God's rest with creation reveal the eschatological significance of God's rest for the creation. When God returns to rest with the creation, the joy and pleasure of God becomes that of the creation; it is the anticipation of the eschatological consummation of nature and history. When, through God's rest, God allows God's self to be affected, to be touched by each of God's creatures, it is the anticipation of our eschatological sensual experience of seeing God face to face (1 Cor 13:12). When God is wholly present in the Sabbath of creation, it is in the anticipation of the eschatological state of God's being all in all (1 Cor 15:28). Therefore, God's rest of creation can be seen as an eschatological anticipation of God's final rest at the end of all history.

According to Moltmann, while creation can be seen as God's revelation of God's works, only the Sabbath is the revelation of God's self, a revelation direct and unmediated.[138] If the Sabbath is the completion of creation and at the same time the revelation of God's reposing existence in God's creation, then "these two elements point beyond the Sabbath itself to a future in which God's creation and his revelation will be one. That is redemption."[139] Therefore, the Sabbath of creation is already the beginning of the kingdom of glory—the hope and the future of all created beings.

This eschatological interpretation of God's Sabbath of creation also places the weekly Sabbath, in commemoration of God's Sabbath of creation, in an eschatological light.[140] When men and women rest from their human works, this becomes a foretaste of the eternal feast of the

---

136. Ibid., 278–80. The emphasis is the author's.
137. Ibid., 280.
138. Ibid., 280, 87.
139. Ibid., 288.
140. This parallels the weekly Sunday celebration of Christ's resurrection as the inauguration of the new creation. I will turn to this subject in chapter 4.

divine glory.¹⁴¹ Therefore, Moltmann also could say that the "Sabbath of God's creation already contains in itself the redemptive mystery of God's indwelling in his creation."¹⁴²

The Form of God's Eschatological Rest: Perichoretic Indwelling in the Creation

Moltmann's concept of the relationship between God and the creation in God's eschatological Sabbath is usually described as panentheism, where God embraces all creation in God's self, without dissolving either God or the creation. But exactly what kind of relationship is supposed here? God's being all in all needs more theological unpacking. In this regard, Moltmann suggests God's *perichoretic* indwelling in the creation is analogous to the three divine Persons lovingly indwelling in one another. This direction is consistent with his overall Trinitarian theology: "If we cease to understand God monotheistically as the one, absolute subject but instead see him in a Trinitarian sense as the unity of the Father, the Son and the Spirit, we can then no longer, either, conceive his relationship to the world he has created as a one-sided relationship of domination."¹⁴³

Moltmann uses *perichoresis* in his Trinitarian theology. He opposes Barth's assertion that there is "in God himself an above and a below, *a prius* and *a posterius*, a superiority and a subordination post-order,"¹⁴⁴ along with Barth's ensuing anthropological assumption of "the ministering body of a ruling soul."¹⁴⁵ Barth's idea that the order of the ruling Father and the obedient Son within the Trinity has an unfailingly correspondent idea of an extra-Trinitarian order of God's rule over the world,¹⁴⁶ and an indestructible order of the ruling soul over the obedient body. Instead of Barth's view of the Trinity and all the ensuing anthro-

---

141. Ibid., 280.
142. Ibid., 280.
143. Ibid., 2. Also: "[T]hinking in relationship and communities is developed out of the doctrine of the Trinity, and is brought to bear on the relation of men and women to God, to other people and to mankind as a whole, as well as on their fellowship with the whole of creation. By taking up panentheistic ideas from the Jewish and the Christian traditions, we shall try to think *ecologically* about God, man and the world in their relationships and indwellings." *Trinity and the Kingdom*, 19.
144. Moltmann, *God in Creation*, 258.
145. Ibid., 252–55.
146. Ibid., 254.

pological assumptions, Moltmann starts with the *perichoretic* indwelling of the Trinity. For Moltmann, the unity of God is achieved through the *perichoretic* union, where there is no command-obedience structure. Moltmann writes:

> [The unity of the triune God] is already to be discovered in the unique, perfect, perichoretic fellowship of the Father, the Son and the Holy Spirit. Following Johannine theology, we are taking as archetype of all the relationships in creation and redemption that correspond to God, the reciprocal *perichoresis* of the Father and the Son and the Spirit (John 17:21).[147]

Here, the understanding of the relationship in God is carried over to the understanding of the relationship between God and the creation; the reciprocal *perichoresis* of the Father and the Son and the Spirit serves as the archetype of all relationships in creation and redemption. Thus, the relationship of the triune God to the creation is not a one-sided relationship of domination; rather it is a complex relationship of fellowship, one of reciprocity. The creative Spirit takes creation into God and God pervades God's creation, allowing the creatures to "exert influence on [God]."[148] In short, in God's rest, God forms a *perichoretic* union with God's creation, a reciprocal relationship originally found in the immanent Trinity.

Some scholars do not feel comfortable with Moltmann's eschatological panentheism. Regarding eschatological perichoretic Sabbath, Brian J. Walsh raises this question:

> Redemption, then, is not primarily a restoration of a covenantal relationship broken in history, but a 're-filling' of that space, an overcoming of God's self-limitation by means of an *annihilatio nihili*. But if this *nihil* was for some reason necessary for the *creatio originalis* then how can the God / creation distinction still be maintained when this *nihil* is vanquished and God is all in all? Is the distiction between pantheism and panentheism only semantic?[149]

Walsh seems to be concerned about the possibility of a taint on God's glory by being too close to the creation. Steven Bouma-Prediger

---

147. Ibid., 258. The emphasis is mine.
148. Ibid., 258.
149. Walsh, "Theology of Hope and the Doctrine of Creation," 75.

takes it a step further, criticising Moltmann's eschatological vision as failing to affirm the creatureliness as well as the otherness of God in spite of Moltmann's claim that the eschatological kingdom does not mean a pantheistic dissolution of creation in God.[150]

I suggest that these kinds of criticisms arise out of an inadequate understanding of *perichoresis*. As its origin in a social Trinity suggests, *perichoresis* does not mean an amalgamation of two different elements but a harmonious dancing of already distinct persons. In fact, the interactions and communications implied in the concept already presuppose the difference in the persons involved. Therefore, just as the same way the three Persons in *perichoresis* are still distinct from each other, the eschatological Sabbath in perichretic union with the creation does not dissolve God and the creation nor taint God's glory nor deny creaturliness; only the alienating distance experienced in history is excluded.

CONCLUSION

In this chapter, I presented the overall framework of Moltmann's ecological theology as a highly eschatologically-oriented concept of the Trinitarian history of God. I first explicated this concept as an ecological cosmology, showing the affinity that Moltmann's ecological theology has with this newly emerging dynamic, non-hierarchical vision of reality. Moltmann's eschatological panentheism and his concept of the Trinitarian history of God not only add a theological dimension to the science-oriented meta-narrative, but also enwrap it by presenting the Trinitarian God as the environment of the cosmogenetic reality. This provides a redemptive *telos* to the ever-evolving cosmic process, making it intelligible to a Christian mind, and eventually rendering the cosmogenetic reality conceivable as God's creation.

Then, I summarised the characteristics of Moltmann's doctrine of the Trinity in terms of social doctrine, Christ-centeredness, and mutual relationship between the immanent and economic Trinity. I presented Moltmann's concept of the Christ-centered Trinitarian history of God in terms of the Trinity in sending and the Trinity in glorification.

Using this as a basis, I presented Moltmann's doctrine of creation with an explication of his panentheistic spatial thinking. In doing this,

---

150. Bouma-Prediger, "Creation as the Home of God," 80.

I made a special effort to reveal the role of the Holy Spirit in His/Her eschatologically bonding, nurturing and guiding connection.

In the last part of the chapter, I presented the consummation of Trinitarian history as the glorification of God in the cosmos using the Trinitarian term *perichoresis*. God finds eschatological rest in the perichoretic indwelling in creation, which is the redemption of all creation. At this point it becomes clear that Moltmann's eschatological panentheism, in terms of the Trinitarian history of God, employs both time and space as God's mission field.

Throughout this chapter, I tried to reveal and emphasise the mystery of the double status—the "already" and the "not yet"—of the creation in relation to God. This tension relates theologically to the concept of an eschatologically-oriented continuing creation and the umbilical connection of the Holy Spirit to the created world. In ecclesiological and missiological thinking, however, this tension constitutes the basis of a congregation coming together for worship. In worship, the oppositional elements of the suffering of the creation and anticipation of new creation are brought together in tension. Then, this tension serves as energy for sending the congregation out into the world for mission.

This tension—under which the whole creation is placed, along with its ecclesiological and missiological connection to Christian worship—does not become clear or concrete enough until it is distilled down to Jesus Christ, who, as our righteousness, holiness and redemption, becomes the foundation for the new creation and thus for an ecological redemption. Only when we start with the Christ event can we conceive of the beginning and the end of the Trinitarian history of God. Only when we ground the ecological hope on Jesus Christ, whom we follow and obey in life and death, does the ecological cosmology transform into an ethic we can concretely relate to in everyday life. That is why I now turn to the cross and resurrection of Jesus Christ.

3

The Cross and Resurrection as the Foundation for Ecological Redemption

ESCHATOLOGICAL CHRISTOLOGY IN THE FRAMEWORK OF TRINITARIAN HISTORY

Paradigms of Christology

MOLTMANN'S METHODOLOGY IN PURSUING his Christology can be summed up as "a Christology that points forward," which significantly complements the two previous christological paradigms, namely, the metaphysical "Christology from above" and the anthropological "Christology from below."[1] Although this eschatological direction of Christology is consistent with Moltmann's earlier books such as *The Theology of Hope* and *The Crucified God*, Moltmann admits that his focus in those books was solely on the resurrection and the cross and thus failed to present an integral Christology that sets "the eschatological history" of Jesus as "the One who will come."[2]

In *The Trinity and the Kingdom of God* and *The Way of Jesus Christ*, however, Moltmann attempts to place Christology in the broader context of a world-encompassing Trinitarian history; in so doing, he bases

---

1. *Way of Jesus Christ*, 3.
2. Ibid., 3–4.

the Christian ecological hope bound with Christ's history in apotheotic assumption into the fellowship of the Trinitarian God. In this scheme, Christ is not a timeless being, nor a complete being but takes historical roles and titles related to His unique mission from the Father and walks His way toward the consummation of the new creation.[3]

This approach to Christology sets Christ's history in a more organic relationship with both the convolutions of the world and its future, and God's Trinitarian history involving the world's history. Therefore, this approach can be rightly called an "integral" approach. As the core and visible promise for the creation, Christology then takes a central place for a Christian ecological hope. This integral Christology set in the context of the Trinitarian history aims to overcome the weaknesses of the two preceding Christologies.

Cosmological Christology of the Patristic Church

First, in claiming that Jesus has a future that He continues to walk toward, Moltmann's integral Christology overcomes the ancient church's metaphysical Christology of two natures that does not consider the history of Jesus as the Jewish Messiah and His eschatological future bound with the future of the creation. In the ancient cosmocentric world, where the individual was always thought of as a member of various concentric communities—e.g., family, the *polis*, and the cosmic community—a human being was also aware of three things: his humanity, that he comes from the earth, and that shares the fate of the earth.[4] In this milieu, the question about existential finitude, transience, and mortality is inevitably linked up with the cosmos in which humans participate and to which humans correspond. Moltmann sees the doctrine of the two natures of Christ, formulated in the patristic church, as corresponding to its cosmocentric worldview. The salvific vision of human participation in the

---

3. "... the rule of Christ, for its part, serves the greater purpose of making room for the kingdom of glory and of preparing for God's indwelling in the new creation, "so that God will be all in all." The lordship of Christ, the risen One, as well as the kingdom of the One who is to come, is in an eschatological sense *provisional*. It is only completed when the universal kingdom is transferred to the Father by the Son. With this transfer the lordship of the Son ends. But it means the consummation of his sonship. It follows from this that all Jesus' titles of sovereignty — Christ, *kyrios*, prophet, priest, king, and so forth — are *provisional* titles, which express Jesus' significance for salvation in time. But the name of Son remains to all eternity." *Trinity and the Kingdom of God*, 92. The emphases are the author's.

4. *Way of Jesus Christ*, 46.

glory of the divine Being, thanks to the eternal Son of God becoming human, therefore, does not exclude the whole cosmos; instead the whole cosmos is drawn into glory with humankind.[5] The theological constitution of Christ corresponding to this kind of soteriology is consequently the constitution of the God-human being, a premise that enables us to substantiate theologically the redemption of humanity and the rest of the creation through him.

Among several impasses of this two-nature Christology, Moltmann argues that the most important one for the current theme is that it was forged in the specific socio-political situation of Roman Empire. Moltmann notes:

> The church's concern for orthodoxy in doctrine and liturgy, and for the exclusion of heresy, was entirely in line with the emperor's concern for a unified imperial religion. . . . In its Christology, Christ was not merely the head of the church. He was also the king of heaven and the pantocrator, thereby legitimating the Christian *imperator* and his empire.[6]

This political *Sitz im Leben* gave patristic Christology two important inclinations. First, as an imperial theology, patristic Christology was a chiliastic triumph, an eternal Christology which sacrificed both a Christology on the road and a Christology beneath the cross.[7] Second, the Orthodox Church lived in this "Christian World" like the soul in the body at the sacrifice of its own bodiliness, that is, a visible community of believers with particular historical and social tasks in the empire such as resistance and transformation.[8] For the patristic Christology, the shift of focus from Easter to Christmas was in a sense inevitable because "the whole of salvation was in fact already implicitly implemented when human nature was assumed by the personally determined divine nature."[9] The importance of the promise in the resurrected body of Christ fades under the light of this *already-salvific* moment of incarnation.

---

5. Ibid., 47. Also, "The wretchedness and the salvation have to do with the very nature of human beings, and together with *their* nature, the nature of the cosmos too, of which they are members." Ibid., 48. The emphasis is the author's.

6. Ibid., 54.

7. Ibid., 55.

8. Ibid., 54–55.

9. Ibid., 51.

There is, however, some important aspects Moltmann cherishes in the patristic christological paradigm: its cosmological horizon and two-nature Christology provide a vision of salvation for the whole cosmos. Moltmann's theology has always carried a strong sense of transience and death as the fundamental threat to the creation lurking in nature and his eschatological vision always involves the resurrection of the cosmos, the annihilation of death, and the new creation. His strong emphasis on the physical dimension of a Christian salvific vision based on the bodily resurrection of Christ also parallels the patristic paradigm.

In spite of this cosmological horizon in the patristic paradigm, however, Moltmann wishes to overcome and complement it with an eschatological christological paradigm that takes into account the unredeemed world and that Christ Himself has a future. Moltmann criticises patristic Christology for its lack of eschatological outlook because it contains neither any real hope for the suffering people nor for the groaning creation.[10]

Anthropological Christology of the Modern Period

Second, Moltmann's integral Christology qualifies a modern anthropological Christology that focuses on human history and the meaning of Christ for that history. The modern anthropological Christology does not consider the violence that "sovereign" humans are wielding on nature; therefore, it cannot address today's ecological concern for, in this theology, Christ exists only in and for history, and has nothing to do with nature.

For the last hundred and fifty years, Christian theology treated the resurrection of Christ in the paradigm of history and disregarded nature. In this paradigm only the eschatological *act in history* of the God who raises from the dead matters; the bodily character of the Christ who died

---

10. Moltmann's political sensitivity combined with his eschatological perspective becomes most evident when he points out the political *Sitz im Leben* of patristic Christology. He argues that the one-sided emphasis on exalted Christ—*pantocrator*—in patristic Christology was entirely in line with the emperor's concern for a unified imperial religion. Accordingly, the eschatological zeal, the hope for liberation, the imagination for a different future become stifled by its christological legitimating of the Christian emperor and his empire — Christ the *pantocrator* in heaven and the Christian emperor in Rome or Byzantium or Moscow. Moltmann beseeches the church "to give up the untimely dream of the *pantocrator* and the imperial church and to turn back to the one crucified, and to live in his discipleship." Ibid., 54–55.

The Cross and Resurrection as the Foundation for Ecological Redemption

and rose again is absent.[11] Of course in this context, "history" means the history of human beings, a progressive history of subjugating the body and nature that has no history.

With the coming of individual, conscious subjects who achieve "scientific and technological civilisation," with the entailing cost of alienation from their own human bodies and natural communities, the identity crisis of human beings is of vital importance in the modern world: what is the human being and what is truly "human"?[12] Moltmann points out that in European and American humanism, a heightened awareness of human dignity and human rights grew up as a response to the increasing subjugation of nature.[13]

As this cultural revolution of Western society continues and a new sense of truth dominated by subjectivity appears, the question of salvation came also to be defined in anthropocentric terms and the old metaphysical, incarnational Christology became irrelevant.[14] Jesus came to be understood not as the God-human being but an anthropological foreground, a perfect prototype of humanity, a true human being, a true image of God which all human beings are supposed to arrive at. The focus in Christology is no longer the incarnation, and the deification of human beings and creation, but Jesus' human sinlessness which was regarded as a divine miracle of love in the moral world.[15] Not surprisingly, this Christology is soteriologically linked to the inward realm of subjectivity, soul, and the human heart.[16]

In conclusion, modern anthropological Christology was a theology for the subjectivity in a divided human being, separated from his/her own body, the community, and the natural world. Therefore, modern anthropological Christology failed to address hope for nature, as did the ancient church's doctrine of the two natures of Christ.[17] The resurrected *body* of Christ has no salvific meaning and carries no promise for this theology.

11. Ibid., 47–48, 247.
12. Ibid., 56–57.
13. Ibid., 57.
14. Ibid.
15. Ibid., 57–58.
16. Ibid., 63.
17. Ibid.

"On the Way" Christology

The Christology of the ancient church, mainly because of its *Sitz im Leben*, did not mind the horizontal history of God's coming Reign. Modern anthropological Christology took as its subject the life and historical personality of Jesus, and thus, does not provide any relevant ecological hope for today because the cross and resurrection of Jesus is not only out of purview, it lacks salvific meaning.

To overcome and integrate these two paradigms of Christology, Moltmann pursues an "on the way Christology", or a "Christology in the movement of God's eschatological history."[18] Here, two significant elements are interwoven to develop an ecological Christology: relationality and eschatology. This Christology will be developed in the Trinitarian context, under the concept of the Trinitarian history toward the eschatological union of Trinitarian Persons with the world. Thus, the horizon of this Christology is no less than God's horizon that includes the *cosmic* horizon of the future and God's Trinitarian blissful union, which is larger than the original unity because of the inclusion of the cosmos in its fellowship.

Because the eschatological history of God the creator encompasses both human beings and nature, the Christology in the eschatological history of God prefers to see Jesus as the Christ who is on His way in the world-encompassing history of the Trinitarian God. Here we must admit that the context and promise in the story of Jesus are considerably broader than that of the anthropocentric Christology of our time. This broader christological horizon is necessary for today because the spiritual foundation for a respectful reconciliation with nature begins to form in the perception of the cosmic Christ and the entailing hope of a cosmic redemption, both in its temporal and spatial dimensions.

Another important aspect of this Christology is to look at Jesus as a social person. Here we consider Christ not only in relation to the other two divine Persons, but also in His identification with, solidarity with, and representation of, the poor and the sick, the God-forsaken sinners, the community of His followers and the whole cosmos. This social

---

18. This image emphasises the anticipatory and provisional nature of an eschatological Christology. It inevitably takes a narrative form and describes Jesus as the "Christ-in-his-becoming". Ibid., 33, 70.

aspect of Jesus plays a significant role in understanding the cross and resurrection in the context of the Trinitarian assumption of the world.[19]

In this way, Moltmann aims to develop a Christology in the Trinitarian history of God, considering Jesus in His relation to the future (eschatological), in relation to His community with others (ecological) and in relation to God (Trinitarian). Both cosmocentricity (patristic two-nature Christology) and anthropocentricity (modern Jesuology) in previous paradigms are not discarded, but incorporated into a theocentric narrative based on the concrete historical narrative of Christ's passion. Accordingly, it contains an ethical outlook that emphasises and seeks the interdependence and reconciliation of humanity and its history on the one hand and the exploited and suffering nature and its history on the other hand.

Although this chapter and this thesis on the whole mainly focuses on the cross and resurrection of Christ in relation to Trinitarian history, it must be noted that Moltmann's "on the way" Christology consistently manifests eschatological direction and cosmic breadth in his close attention to Jesus' earthly ministry. Jesus' ministry of healings and exorcisms, for example, points to a holistic salvation that includes bodily and cosmic dimensions and this anticipates the new creation of all things.[20]

Christology in the Trinitarian History of God

The Interweaving of Christology and the Doctrine of the Trinity

Moltmann's theology has always been strongly Christ-centred because he believes that the history of Christ's passion is at the heart of the Christian proclamation of God. This means his overall theology, including the doctrine of the Trinity, revolves around the axis of Jesus Christ, always flowing into and out of the living Christ, who is on His way to the eschatological consummation. In Moltmann's view, the Christian faith is wakened in us "when the passion of Christ becomes present to us through word and sacrament."[21] Furthermore, "God him-

---

19. Ibid., 71. This point will be dealt with in more detail in the section 2.2. "The dimensions of the death of Jesus" in this chapter.

20. Ibid., 105–10. In fact, the place of chapter 3 (The Messianic Mission of Christ) in *Way of Jesus Christ* is intended to show the sequential nature of narrative form Christology and its eschatological direction. Ibid., 71.

21. *Trinity and the Kingdom*, 21.

self is involved in the history of Christ's passion" so that the Christian believes in God for Christ's sake.[22] This position of Moltmann makes his theology highly christological and yet never christomonistic; instead, it makes Moltmann's theology a *Christ-centered eschatological Trinitarian theology*.

This interweaving of Christ's passion and God's dealings with the world demands a Trinitarian understanding of Christ's story, especially the cross event. Christ's history with the cross event at its core cannot be separated from the Trinitarian history of God or understood as an emergency measure to cope with human sin; instead, it should be placed in the broader narrative of the Trinitarian embrace of creation. Here, the doctrine of the Trinity has Christology and the understanding of the cross event as its premise.

Conversely, the Christology that is presupposed for the doctrine of the Trinity must be an open Christology. Moltmann maintains that it should be "open for perception of the creation of the world through the Father of Jesus Christ, and open for perception of the transfiguration of the world through the Holy Spirit, who proceeds from the Father of Jesus, the Son."[23] In this connection, Moltmann's Christology always binds Jesus Christ in a strong relationship with the other two divine Persons and with the rest of creation. Jesus Christ is not an isolated ready-made "subjectivity." Instead, being a Person in the making, Christ is on His way to the fullest expression of Sonship, both in a Trinitarian relationship to God the Father and the Holy Spirit and in His social relationship with other people and with nature. Here again, we can identify the combination of an eschatological process and relationality in Moltmann's Christology.

Because of this inter-relationship of Christology and his Trinitarian doctrine, Moltmann's Christology is considered in the context of the Trinitarian history. In other words, at the core of his doctrine of the Trinity is Christ's passion and resurrection story.[24] Therefore, Moltmann emphasises the inseparable relationship between the immanent and the economic Trinity, appropriating Barth's idea of retroactive relationship of the cross in time into the eternal being of God.[25]

22. Ibid.
23. Ibid., 97.
24. *Crucified God*, 202.
25. "It is only in his account of Christ's death on the cross that Barth breaks through

The Cross and Resurrection as the Foundation for Ecological Redemption 75

Eschatological-Cosmic Perspective on the Cross and Resurrection in the Trinitarian History of God

The close tie between Moltmann's Christology and his doctrine of the Trinity demands a Trinitarian interpretation of the Christ event, that is, an interpretation of the cross and resurrection in the context of God's historical movement toward world-embracing redemption. In fact, from his early theology, Moltmann viewed the cross and resurrection as a decisive eschatological event in which the transition from the old aeon to the new happens. In his full-fledged ecological Christology in *The Way of Jesus Christ*, Moltmann develops this idea even further in eschatological and Trinitarian sensitivity.[26]

First he sets the cross of Christ within the Jewish apocalyptic horizon.[27] While the whole creation has been under suffering—and the particular suffering of Christ can be seen as part of the sufferings of Israel and God's prophets—the sufferings are not just Christ's own but through them, "the end-time sufferings of the whole world are anticipated and vicariously experienced."[28] In this context, Christ suffers the great apocalyptic dying, the death of all things, bearing the sufferings of the victims of this present time, who are the weak, the poor, the sick, the oppressed, the children, and nature.[29]

Moltmann appropriates this apocalyptic thinking in the context of nuclear and ecological end-time possibilities. In this apocalyptic understanding, the sufferings of Christ are seen as the unprecedented suffering

---

the unilinear view of correspondence, which thinks of it from above to below, from within to without. Christ's death on the cross acts from below upwards, from without inwards, out of time back into the divine eternity. . . . The meaning of the cross of the Son on Golgotha reaches right into the heart of the immanent Trinity." Ibid., 159.

26. The ecological interpretation of the cross and resurrection in *Way of Jesus Christ* is a relatively new development from his early Christology in *Crucified God*. But this is not an alien insertion because the ecological theme has been consistently growing since his early trilogy and the inclusion of the world in the split Trinitarian relationship was already there in his early theology.

27. Moltmann's chapter 4, "The Apocalyptic Sufferings of Christ" in *Way of Jesus Christ*, 151.

28. Ibid., 155.

29. "In the struggle for power which is the trademark of 'this world,' the weak suffer most, the oppressed are sacrificed first of all, the children are the first to die. In the struggle for wealth, people destroy the creatures that are weaker than themselves. Nature dies her dumb death first of all, and the death of the human race follows." Ibid., 157.

in the end-time. But at the same time, because they are the birth pangs of the new creation, the dawn of a new era is expected. Thus, Moltmann views the whole creation, including humanity, its history, and nature, as having gone through the transition from old to new creation because of the eschatological transition of Jesus Christ, who assumed the whole creation in His vicarious apocalyptic sufferings on the cross.[30]

It must be noted here that the epistemological order of these two phases of a single event is backward: because Jesus' resurrection was perceived as the anticipation of the new creation, His passion was a vicarious suffering of the end-time sufferings. This is why the resurrection of Christ is the epistemological foundation for cosmic hope and the cross as the ontological foundation for ecological redemption.[31]

Now out of this eschatological-cosmic interpretation of the cross and resurrection, the ecological significance of Jesus Christ emerges. Because Christ died not only for human sin but He died in solidarity with all living things, consequently the resurrection is good news not only for humanity but also for the whole creation. In an eschatological-cosmic perspective on the cross and resurrection, something much larger looms than an anthropocentric significance of the cross and resurrection of forgiveness of sin: the annihilation of the death and transience in nature. In summary, Moltmann remarks:

> [Christ] did not merely die the violent death which belongs to human history. He also died the tragic death of nature. If his resurrection is the death of death, then it is also the beginning of the annihilation of death in history, and the beginning of the annihilation of death in nature. It is therefore the beginning of the raising of the dead *and* the beginning of the transfiguration of the mortal life of the first creation in the creation that is new and eternal.[32]

Furthermore, if Christ's cross and resurrection is interpreted in this eschatological-cosmic perspective, Christology stands at the center of a Christian appreciation of nature's intrinsic value and the hope of its

---

30. Moltmann's chapter 5, "The Eschatological Resurrection of Christ" in *Way of Jesus Christ*, 213–73.

31. "The epistemological foundation is the resurrection of Jesus. . . . The ontological foundation for cosmic Christology is Jesus' death on the cross." McWilliams, "Christic Paradigm and Cosmic Christ," 348.

32. Ibid., 253. The emphasis is the author's.

ecological redemption. Christ's vicarious sufferings and His anticipatory transition as the first-fruit of the dead assure an infinite value of all things in the created world. Moltmann argues:

> If Christ has died not merely for the reconciliation of human beings, but for the reconciliation of all other creatures too, then every created being enjoys infinite value in God's sight, and has its own right to live; this is not true of human beings alone. If according to the Christian view the uninfringeable dignity of human beings is based on the fact that 'Christ died for them,' then this must also be said of the dignity of all other living things. And it is this that provides the foundation for an all-embracing reverence for life.[33]

Now we can turn to see in more detail how the cross and the resurrection can be interpreted in the cosmic horizon and in relation to Trinitarian history.

THE CROSS AS THE ONTOLOGICAL FOUNDATION FOR ECOLOGICAL REDEMPTION

The Cross as the Division and Unity in the Trinity

For Moltmann, the cross is always at the center of the Trinity; the cross reveals the heart of the eternal triune God and it has made an indelible retroactive imprint on the divine life of the Trinity. For a deeper understanding of the meaning of the cross, not for the world but for the Trinitarian God, Moltmann returns to a theology of surrender in the New Testament.

He pays close attention to the word *paradidonai* (deliver up, betray, hand over, cast off) to clarify the meaning of the death of Jesus to God.[34] When this word is applied to the death of Jesus on the cross, it denotes unmistakably a negative significance that God has forsaken and cast off Jesus. But when Paul views the God-forsakenness of Jesus not only in historical perspective but in light of Jesus' resurrection, he radically reverses the meaning of "delivered up" and this word begins to take on

---

33. Ibid., 256.

34. Usages of this word, for example, in portraying Judas as "the betrayer" or in divine judgment on human sin (Rom 1:18ff.) show the negative undertone of this word. *Way of Jesus Christ*, 172.

a salvific meaning. Because of the resurrection and His presence in the Spirit, Paul concluded that the Father has forsaken "his own Son ... for us";[35] the giving up of the Son by the Father seen in light of the resurrection constitutes a Trinitarian salvific act.

> The Father forsakes the Son "for us"—that is, he allows him to die so that he may become the Father of all those who are "given up" (Rom 1:18ff.).... The Son is surrendered to this death in order to become the brother and saviour of all the men and women who are condemned and accursed.[36]

The Son, however, was not subjected to a mere passive death. According to Gal 2:20, the Son also "gave himself" for us, choosing a deliberate path of suffering. This conformity between the will of the surrendered Son and the surrendering will of the Father—a profound *community* of will—paradoxically happens at the very widest point of *separation* of the Son from the Father, at the breaking off of the intimate relationship between the Father and the Son, in the instance of the God-forsaken death of God's Son.[37]

At this point, Moltmann finds the Augustinian notion insufficient, that the *opera trinitatis ad extra* (The works of the Trinity on the outside) are undivided (*indivisa*) and the *opera trinitatis ad intra* (The works of the Trinity on the inside) are divided (*divisa*). For the cross event (*opera trinitatis ad extra*) can only be understood in terms that are *divided* and *differentiated*.[38] For Moltmann, the split of the Trinity in the outward creation of salvation corresponds to, and retroactively acts on, the intra-Trinitarian suffering of disastrous separation, reaffirming his emphasis on the close inseparable relationship between the immanent Trinity and the economic Trinity.

---

35. Here, Moltmann believes the Trinitarian thinking starts when we look back at the cross in light of resurrection. The death of Jesus as a false messiah, a prophet, or as one of the righteous men of the people would be nothing special, nor would it involve mystery, as many such deaths occurred before Jesus. Only when we go back from the resurrection to the cross, from God's raising of Jesus as *kyrios* and God's Son to the God-forsaken death on the cross, a theological contradiction emerges from which the Trinitarianism grows up. Ibid., 170–71.

36. Ibid., 173.

37. "This event contains community between Jesus and his Father in separation, and separation in community." *Crucified God*, 244.

38. *Trinity and the Kingdom*, 160.

The Cross and Resurrection as the Foundation for Ecological Redemption

Another important element in this community of the will of the Father and the Son to surrender lies in the fact that the surrender is made "through the eternal Spirit" (Heb 9:14). Moltmann highlights the binding role of the Holy Spirit, emphasising that the cross is a Trinitarian event:

> The Holy Spirit is the bond in the division, forging the link between the originally lived unity, and the division between the Father and the Son experienced on the cross. . . . [T]he Spirit who was Jesus' active power now becomes his suffering power. . . . The Spirit is the divine subject of Jesus' life-history; and the Spirit is the divine subject of Jesus' passion history.[39]

In the Spirit, the Father and the Son are connected with indestructible fellowship, unity, community, a soteriologically inviting space for the cosmos.[40] Thus, the cross is a Father-Son event of division and unity in the Spirit.

The Dimensions of the Death of Jesus

To probe the salvific meaning of Jesus' death in the division and unity of the Trinity, Moltmann asks the theological meaning of His death: "What death did Jesus die?" In other words, "Who was he in this death of his?"[41] Moltmann enumerates five dimensions of Jesus' death: 1) the death of the messiah, 2) the death of God's child, 3) the death of the Jew, 4) the death of the slave, and 5) the death of the living One.[42]

Consistent with Moltmann's social and eschatological emphasis in his Christology, Jesus dies in solidarity with the poor, and the people of Israel. Moltmann also connects the Trinitarian involvement in Jesus' death as the death of God's child. But the link with his ecological concern is emphatically the death of the living One.

For Moltmann, when Jesus dies the death of the living One, He participates in the fate of all living things: in His physical struggle at death, in His desperate gasping for breath, in the increasingly difficult

---

39. *Way of Jesus Christ*, 174.

40. This is what is meant when Moltmann says "Through the sending, the fellowship of the Father and the Son becomes so all-embracing that men and women are taken into it, so that in that fellowship they may participate in Jesus' sonship and call on the Father in the Spirit." *Trinity and the Kingdom*, 75.

41. *Way of Jesus Christ*, 160.

42. Ibid., 164–69.

panting, in His burning thirst and crushing headache; this He shares with all dying animals. Here, it is important to note that Moltmann no longer considers Jesus' death as a uniquely human tragedy; Moltmann views death as the fate of *everything* that lives, pointing to the tragedy embedded in the natural process of life in creation. Therefore, Jesus died the tragic death of nature that is intrinsic in the creation, not yet fully indwelled by the life-giving Spirit.

In connection with the death and transience in nature, Moltmann defies the traditional dogma that views sin as the single cause of death both in humanity and beyond it. Paul and Augustine teach that the wages of sin is death; Jesus, the sinless Son of God cannot have died His own death, but only a vicarious, compassion-filled death. This perspective reduces Jesus' death to a human death, not the universal death found everywhere in the created order. Instead of this view, Moltmann opts for Schleiermacher's view that mortality belongs to creaturely finitude, something that is intrinsic in nature, instead of being a punishment of sin. The wages of sin could well be "the fear of death but not physical death itself."[43] Adopting this view of death in nature does not exclude the vicarious nature of Jesus' death. Instead, in Moltmann's view, "Jesus could therefore die 'the accursed death of sin' vicariously for all sinners only because he was *sinless and mortal at the same time*."[44]

Once we accept Jesus' death not only as a vicarious death for sinners but also as His own natural death to which as part of the created order He was subject, we can see His death as being in solidarity with the whole creation, that sighs under the permanent threat to transience and yearns for deliverance from that threat.[45] To perceive this dimension of Jesus' death as being in solidarity with the whole creation comes from His resurrection, that which overcame the "natural" process of decaying. Conversely, this clear identification of Jesus' death as the end of all finite living things that spring from the physical and natural world forms the basis of our hope for the whole cosmos in the resurrection of Jesus Christ.

---

43. Ibid., 169.

44. Ibid. Emphases are mine.

45. This view of Moltmann on nature stresses nature's need for redemption and runs counter to the ecofeminist trend of "christification of creation," such as seen in Sally McFague. McWilliams, "Christic Paradigm and Cosmic Christ: Ecological Christology in the Theologies of Sallie McFague and Jürgen Moltmann," 351.

The Cross and Resurrection as the Foundation for Ecological Redemption

The Embrace of the Cosmos in the Trinitarian Event of the Cross

Up to now, we have viewed the cross as a Trinitarian event of division and unity. We also have observed Moltmann's emphasis on Jesus' death in solidarity with the sighing creation. When we put these two ideas together, we must conclude that the creation is also included in the Trinitarian event of the cross.

What is important here in relation to Moltmann's thinking of the Trinitarian God in the event of the cross is his *spatial* thinking; the Trinitarian relationship becomes the event or space that encompasses all of creation's time and space. Moltmann expresses the cross event as an opening of a profound cosmic-scale space which includes the creation:

> When Paul uses the *paredoken* formula, he always speaks of Jesus, of the "Son of God," never of Christ or of *Kyrios*, as if he wished to say that in the giving up and the abandoning of Jesus by the Father a cleft opens up in God which reaches so deeply that through it every cleft of sin and of judgment between God and humankind can be embraced and healed.[46]

In this "bifurcation" in God, "the cross stands between the Father and the Son in all the harshness of its forsakenness,"[47] producing a very deep and very wide space in God.

It is, however, of utmost importance here to note that this new *space* in God made possible by the self-distinction in God at Golgotha functions soteriologically: "The Trinitarian self-distinction of God in the death of the Son on the cross is so deep and so broad that all those lost and abandoned will find a *place* in God."[48] In this sense, Moltmann argues that "the concrete 'history of God' in the death of Jesus on the cross on Golgotha therefore contains within itself all the depths and abysses of human history."[49]

Moltmann takes one more step in his soteriological articulation of the cross in spatial terms. In the event of the cross the creation is not only *contained* in the Trinity but also begins to be *permeated* by the Trinity. Through this Trinitarian event of the cross, the Father reaches

---

46. Moltmann, "Christian Doctrine of the Trinity" in *Jewish Monotheism and Christian Trinitarian Doctrine*, 53.
47. *Crucified God*, 246.
48. "Christian Doctrine of the Trinity," 53. The emphasis is mine.
49. *Crucified God*, 246.

every corner of His creation through the Son, achieving omnipresence, because the Son was cast off into the depth of God-forsaken hell, into loneliness and annihilation. Through this Trinitarian event of the cross, we perceive that the Father becomes the Father and Creator of everything in the creation, because the Son has become the brother of all the Godless.[50] In Moltmann's words, in the cross event, "God is not only revealed in history but also opened up to the experience of history,"[51] suffering every suffering of the creation and enjoying the conversion of humankind and the liberation of nature in the parallel and interwoven missionary history of Christ and the Holy Spirit. Moltmann truly sees in this Trinitarian event of the cross the beginning of the language of the kingdom of God, where "God will be all in all."[52]

This containment and permeation of the cosmos by the Trinity in the event of the cross helps us to expect the extent and degree of the transformative redemption that will be revealed through the resurrection of Christ and manifestation of the Holy Spirit: the extent and degree can be no less than the eschatological cosmic perichoretic indwelling of the Trinitarian God. Of course, the understanding of the cross as the ontological foundation for ecological redemption comes from a backward inference from the resurrection and the presence of the Holy Spirit. Nevertheless, this spatial comprehension of the cross as the Trinitarian event of division and unity provides a clearer understanding of the cosmic dimension of the hope that Christ's resurrection anticipatorily embodies and radiates.

---

50. "The Father, who sends his Son into all the depths and hells of God-forsakenness, loneliness and annihilation, is in his Son everywhere among those who are his, so that he has become omnipresent. With the surrender of the Son he gives 'everything', and 'nothing' can separate us from him." *Way of Jesus Christ*, 174.

51. "Christian Doctrine of the Trinity," 54.

52. *Trinity and the Kingdom*, 82; *Way of Jesus Christ*, 174.

THE RESURRECTION AS THE EPISTEMOLOGICAL FOUNDATION FOR THE COSMIC HOPE

The Cosmic Significance of the Resurrection of Christ

The Eschatological Character of Christ's Resurrection

For Moltmann, the suffering of Jesus is an apocalyptic suffering and it anticipates the tribulations and assailment as birth pangs of the end-times.[53] Furthermore, Moltmann sees Jesus' death not only as a death by human violence (and thus a death in solidarity with all who suffer violence) but also a death of all living things, a tragic death of nature inherent in the created world. This perspective on the cross corresponds to his concept of Jesus' resurrection as the beginning of the death of death, and the beginning of the transfiguration of the mortal life in creation.[54]

To show the eschatological character of Christ's resurrection that involves the future of the whole cosmos, Moltmann discusses the eschatological structure of Easter appearances. The "seeing" in the Easter accounts is used as a revelation formula. This "seeing," however, is not a recognition of something which is always present and everyone can see with careful attention and is open to testing. Rather, this seeing is a vision, a "seeing of something which someone is permitted to see in a particular way."[55] According to Moltmann, the structure of the Easter "seeing" takes the form of "the messianic pre-reflection of what is in the future, and of the apocalyptic anticipation of what is to come."[56] In other words, "Christ appears to the people concerned in light of the future which cannot otherwise be perceived in the world as yet."[57]

Here, the epistemological character of the Easter event becomes clear. The Easter appearances had an eschatological structure and on the basis of what they themselves saw, the disciples took up an apocalyptic symbol of hope (the raising of Jesus from the dead) that makes sense of the contradictory experiences of the disciples—crucified in shame and seen in glory. Therefore, for the disciples, the day of Christ's resurrection was perceived as the first day of the new creation. It was in this cosmic

---

53. Ibid., 153.
54. *Way of Jesus Christ.*, 253.
55. *Trinity and the Kingdom*, 84.
56. Ibid.
57. Ibid., 85.

dimension, Moltmann argues, that Christ's post-Easter appearances were grasped in the New Testament. Moltmann remarks, "Paul makes this clear in II Cor. 4.6, when he traces back 'the glory of God in the face of Jesus Christ' to the Creator who on the first day of creation 'let light shine out of darkness.'"[58] This cosmic perspective on the salvific significance of Christ's death and resurrection is the basis for the cosmic hope for salvation.

The Ecological Significance of the Bodily Nature of Christ's Resurrection

In the eschatological resurrection of Christ with a cosmic implication, the bodily nature of Christ's resurrection has a pivotal soteriological importance. Instead of the salvation of the spirit from the body or from the material world, in accordance with the bodily resurrection of Christ, Moltmann holds that salvation involves the whole human being—both in his/her spiritual and bodily dimensions—and the whole world—both in its material and spiritual dimensions. For Moltmann, human salvation is inextricably linked with the salvation of the whole cosmos due to a human structure (a *perichoretic* union of body and spirit) and is conceivable only under the premise of a physical and material redemption of the cosmos. In this connection, Moltmann discusses the relationship between spirit and body, nature and history to establish the cosmic extent and degree of salvific efficacy of the Christ event.

Spirit vs. Body

Against the prevailing Western culture that values transcendental subjectivity—with its power of cognition and will, and its downgrading of the body to a mere possession—Moltmann regards the likeness in which God created human beings as their whole bodily existence. In this whole bodily existence, the body is neither subordinate nor inferior to the soul.[59] Rather, the body and the spirit are in *perichoretic* union, a relationship of mutual penetration and differentiated unity.[60] Thus, there is neither primacy of the spirit over the body or the primacy of the body over the spirit. The body and the spirit are differentiated but inseparable partners forming one unity. This *perichoretic* unity of the body and soul

---

58. Ibid.
59. Ibid., 265–266.
60. Ibid., 258–259.

The Cross and Resurrection as the Foundation for Ecological Redemption 85

engages the whole human being with nature. As long as humans have a bodily existence, and thus participate in nature, the whole human being participates in nature. That is why human salvation cannot be imagined without the salvation of nature.[61]

Nature vs. History

For Moltmann, the separation of nature and history is a result of a dubious modern enterprise of sovereignty over nature and body, in which history, *vis-à-vis* nature, becomes the great paradigm of the modern world. Moltmann notes that behind this dichotomy of history and nature lies the same problematic logic of mind and body:

> Like history and nature, mind and nature were also defined over against one another: history is the field of freedom, nature the field of necessity. The mind is something quite separate from nature, because it can know and dominate nature; and nature is separated from mind because it can be subjected by the mind.[62]

But the human structure of a psychosomatic unity signals the untenability of the paradigm of history as an all-accommodating model for the perception of reality as a whole.[63] Therefore, like the case of body and mind, history and nature must not be defined over against one another because, just as mind or spirit is embedded in one's bodily nature, human history is wholly embedded in the natural conditions; nature provides the framework of history and history works in, and is totally dependent on, that framework. Therefore, history and nature should be seen as mutually dependent, not with one subject to another. Nature is not aligned with human history. Rather, "Human history is consummated in 'the resurrection of nature,' because only in and through that is a 'deliverance' of human life conceivable."[64]

Bodily Nature of Eternal Life

For Moltmann, eternal life is doubtlessly a life in a body, not without it. In this context, he criticizes the modernized ecumenical version of

61. This theme is picked up again in *Coming of God*, 65–66.

62. Ibid., 246.

63. Moltmann concisely puts it this way: "In the embodiment of a human being his nature and his history coincide.... In his embodiment Christ suffered both the 'historical' and the 'natural' torments of death...." Ibid., 256.

64. Ibid., 254.

the Apostles' Creed that has replaced "the resurrection of the body" with "the resurrection of the dead" as "a modernisation which tries to banish from Christian eschatology its offensive fleshliness, and by doing so makes its hope a cloudy one."[65] But because of the inseparable *perichoretic* unity of body and soul that comprises a whole human being, there can be simply no resurrection at all without the resurrection of the body. By the same token, the identity God remembers about the dead is also a somatic identity because personal identity is found in a person's life history.[66]

The assertion of the bodily nature of eternal life also affirms the material of which earthly things are composed and in which the body also participates. Moltmann asks, "Will human needs and human dependence on food, air, climate and so forth be abolished? . . . Will human sexuality be abolished as well . . . ?"[67] To make these assumptions is Gnostic rather than Christian, because they do not go along with the hope of bodily resurrection. In this connection, Moltmann emphasises the continuity of the old and new creation:

> If we have to assume this, then it is not this creation which is going to be created anew, for in place of the human being who is created male and female there will be a different being altogether, and "the second creation" will displace the first. But the eschatological new creation of this creation must surely presuppose *this whole* creation. For something new will not *take the place* of the old; it is *this same "old" itself* which is going to be created anew (I Cor. 15.39–42).[68]

Therefore only guilt, grief, sin, suffering and mortality are going to be removed from the created nature of eternal life, not human dependence on nature or human sexuality.

The Resurrected Body of Christ as an Embodied Promise of the Deification of *ta panta*

The cosmic significance of Christ's resurrection lies in the fact that this resurrection was not an isolated incident in the past, with no impact on

---

65. *Way of Jesus Christ*, 259. Here, the word "modernization" points to the deserting of the hope for flesh in favor of "personhood" which is typical modern thinking.

66. Ibid., 260–61. Also, *Coming of God*, 66.

67. *Way of Jesus Christ*, 262.

68. Ibid. Emphasis is the author's.

The Cross and Resurrection as the Foundation for Ecological Redemption 87

the present, but is the beginning of the transformation of everything (*ta panta*) toward the new creation. At this point, the bodily nature of Christ's resurrection, as is witnessed in the New Testament, becomes important. Citing Moltmann, "Embodiment is the existential point of intersection between history and nature in human beings."[69] As the resurrection was bodily, it should not be seen only as an act of God in history, but also as the beginning of a new creation of the *old* creation in its bodily, fleshly, and earthly dimensions. Through this event, the body of Christ that went through the transition becomes the model, the prototype, and the law of the new creation; it is exactly in this sense that the resurrected body of Christ becomes an *embodied promise* toward the whole creation that has not yet been resurrected. As an embodied promise, it must be recognized and appreciated as the single focal point that anticipated the eschatological eternal life in the power of the Spirit.

What impact does an embodied promise have on the present life of the unredeemed world? On the one hand, it functions as the basis of hope for the whole groaning creation. If Christ, who died in solidarity not only with the human victims of history but with all the living things in nature, went through the transition, the whole creation (*ta panta*) surely can hope that it will also be transformed after Christ. On the other hand, as an embodied promise, the resurrected body should be working as the power of a new creation right now, in and through the body, that is, in the *material* dimension. In this context, Moltmann affirms Orthodox theology:

> [The raised body of Christ] is the prototype of the glorified body. Consequently, a transfiguring efficacy emanates from it. It is wholly and entirely permeated by the life-giving Spirit. It therefore radiates the Spirit which already gives life *here and now*. It stands in light of God's perfection, so "from the raised body of the Lord streams a boundless ocean of light."[70]

If the living Christ in His bodily-resurrected person emanates a transforming light of new creation into the winter of creation, we could say that Christ's resurrection is the epicenter of ecological hope that everything will be transformed and redeemed into the springtime of creation. In this sense the resurrected Christ is the cosmic Christ. Moltmann

---

69. Ibid., xvi.
70. Ibid., 258. The emphasis is mine.

draws on the cosmic Christology of Colossians which holds that Christ died to reconcile *everything* whether on earth or in heaven (1: 20) and through the power of resurrection He became "the first-born from the dead" and was therefore revealed as "the first-born of all creation" (1:15, 18). The pivotal importance in this cosmic Christology of Colossians is the bodily nature of this cosmic Christ and His resurrection. In this cosmic perspective Moltmann notes that Christ died not only the death of sinners but the death of all the living (1:20), which should include the angels and the beasts, and that the whole fullness of the Godhead dwells in His resurrected body (2:9), from which "this *Shekinah* overflows into the new fellowship of creation."[71]

This interpretation is a reaffirmation of the "deification (*theosis*)" of the cosmos upheld by Orthodox theology. Moltmann draws on Gregory of Nyssa, who describes the resurrection of Christ as letting "everything that with him lay in the dust now rise with him as life and resurrection and sunrise and daybreak and the light of day for those who sit in darkness and the shadow of death."[72]

Rebirth of Jesus in the Spirit

Transition of Jesus as the Beginning of the Eschatological Resurrection of the Cosmos

Moltmann suggests that we see the terms indicating Christ's resurrection—raising, resurrection, making alive, transfiguration, transformation—as describing a *transition*. This transition is not from a bodily existence to a spiritual one; rather, the transition flows from the bodily existence of an old aeon to a new creation. Thus this transition is a progression, a passage "from sleep to waking, from defeat to walking with the head held high, from death to life, from shame to honour . . . from human violation to divine glory, from mortal existence to immortal divine being."[73]

Two things are important about this transition. First, it should be seen as the efficaciousness of the Spirit *in the body*. This transition was made possible by the life-giving Spirit, who transformed Christ's mortal body by wholly and entirely permeating it, giving Christ a rebirth in the

---

71. Ibid., 255.
72. Ibid.
73. Ibid., 257.

Spirit.[74] The person who is wholly and entirely seized and pervaded by the living power of the life-giving divine Spirit—that is, the power of the new creation—becomes immortalized, because death loses its power over him/her.

Second, Moltmann used the term "transition" to emphasize *the continuity* of the old and new forms of the body. In the faithfulness of the Creator, this transition is not from nothing into something, or from something into nothing; instead, this is the old creation—with all its fleshly material, its earthy dimension—being reshaped, reformed and transformed by the power of the Spirit into an immortal order.[75] Thus the term transition connotes Moltmann's insistence that created life is a bodily life both in the old and new creation.

Here, an investigation of Moltmann's concept of new creation in terms of continuity and discontinuity is in order. How new is the new creation for Moltmann? Here, Moltmann is legitimately criticised for not always being clear and sounding contradictory.[76] Schuurman criticises Moltmann as proposing a world-annihilating eschatology on the basis of Motlamnn's expression of the new creation as *creatio ex nihilo*.[77] Surely Moltmann emphasises the newness of the eschaton vis-à-vis the future of the creation as the development of the potentialities already present in it. In Moltmann's words, "The new creation does not emerge out of the restoration of the old creation; it follows from creation's end."[78] On the basis of this emphasis, Schuurman suggests Moltmann's eschaton implies a kind of annihilation of creation. I believe that Moltmann's application of the term *creatio ex nihilo* to the new creation was a mistake. Perhaps his theological intention was to emphasize the *novum* of eschaton, shunning the protological inclination in the idea of *restitutio integrum*.

Later, when he began to pay fuller attention to the ecological issues, he realises the unintended world-denying nuances of this expression and attempts to strike a balance between the newness of the new creation with the Creator's faithfulness to the creation, between the discontinuity and the continuity. This turning of his to an ecological awareness

---

74. *Spirit of Life*, 146–47.
75. Ibid., 150.
76. Walsh, "Theology of Hope and the Doctrine of Creation," 56–58.
77. Schuurman, "Creation, Eschaton, and Ethics," 49.
78. *Future of Creation*, 164.

and his emphasis on the continuity between the present creation and the new creation is present before *God in Creation*. In *The Trinity and the Kingdom*, Moltmann offers a more comprehensive presentation of Trinitarian history, which rightly encompasses the whole of creation and especially emphasizes the eschatological glorification process by the Spirit who was sent from the risen Christ. By doing this, he already corrects his earlier overemphasis on the discontinuity. The perception of the anticipatory presence and the "liberating and perfecting" works of the eschatological Holy Spirit in the present world is especially important here.[79] This realization prompts Moltmann to say, "This glorification of the Father through the Son in the Spirit is the *consummation* of creation. . . . People and things are therefore gathered into the Trinitarian glorification of the Son and the Father through the Spirit. In this way they are also united with God and in God himself."[80] Furthermore, in *God in Creation*, Moltmann reaches the point of declaring that the three types of creation (creation in the beginning, continuous creation, and new creation) are not disjunctive or disparate but "a meaningfully coherent process."[81] This remark is surely a complete departure from his earlier position of new creation as *creatio ex nihilo*. The *transition* to describe the resurrection (radical newness) of the crucified One (radical faithfulness to the creation) as the inaugural event of the new creation suggests this fine-tuning of direction.

The Eschatological Task of the Holy Spirit

The Holy Spirit was the means by which God raised Jesus from the dead. However, because Jesus was raised into the coming glory of the Father, to the innermost being of God the Father, at the right hand of God the Father, being so near to God the Father, now He participates in God the Father's sending of the Spirit, thus becoming the "life-giving spirit" (1 Cor. 15:45).[82] At this critical point of the Christ event, the inter-Trinitarian roles are changed:

---

79. *Trinity and the Kingdom*, 124–25.
80. Ibid., 126. Emphasis is mine.
81. *God in Creation*, 55.
82. Moltmann chooses "through" instead of "by" to express the Son's sending of the Spirit because the Son participates in the Father's sending; in this way, Moltmann eschews a strict *filioque* thinking of double origin of the Spirit.

The Cross and Resurrection as the Foundation for Ecological Redemption 91

> Whereas in the sending, in the surrender and in the resurrection, the Spirit acts on Christ and Christ lives from the works of the creative Spirit, now the relationship is reversed: the risen Christ sends the Spirit; he is himself present in the life-giving Spirit; and through the Spirit's energies—the *charismata*—he acts on men and women.[83]

Because the Spirit is now sent through the risen Christ, whose resurrection is the eschatological anticipation of cosmic redemption, the Spirit receives an eschatological task. Moltmann reminds us that in the whole of the New Testament the Spirit is understood eschatologically, as the power of the new creation, the power of the resurrection, and the pledge of glory to come.[84]

If we speak of the eschatological task of the Holy Spirit in terms of reality and hope, it brings the eschatological future into the present unredeemed reality. The church in the New Testament had a keen sense of the dual reality of anticipated hope alongside the grim day to day experience. But the tension the Holy Spirit brings into the present is broader than the church, or even humanity. "The history of the creative Spirit embraces human history and natural history."[85] Accordingly, the task of the Holy Spirit is much broader in scope than the church's and humanity's concern; no less than the cosmic renewal is embraced within His/Her horizon. Therefore the Holy Spirit has an ever-widening, concentric expansion of mission:

> His present efficacy is the rebirth of men and women. His future goal is the raising up of the kingdom of glory. His activity is experienced inwardly, in the heart; but it points ahead into what is outward and public. He lays hold on the soul, but will only find rest when he gives life to mortal bodies (Rom 8:11).[86]

In this eschatological mission from the heart to the physical, from inward to outward, from the present to the future, the Spirit carries the name and character of Christ, reminding us of him, making Him present in the word and sacrament, and glorifying Him by renewing the cosmos into a new creation after the model of Christ.

---

83. *Trinity and the Kingdom*, 89.
84. Ibid., 89.
85. *Church in the Power of the Spirit*, 34.
86. Ibid.

Trinitarian Doxology in Light of the Resurrection

Moltmann has emphasised the mutual relationship between the immanent and economic Trinity by way of the retroactive effect of the event of the cross.

> On the cross, God creates salvation outwardly for his whole creation and at the same time suffers this disaster of the whole world inwardly in himself. From the foundation of the world, the *opera trinitatis ad extra* correspond to the *passions trinitatis ad intra*.[87]

But if the pain of the cross determines the inner life of the triune God from eternity to eternity, then the same must be true of Christ's resurrection and the entailing eschatological gathering of the creation in Christ through the Spirit, because the resurrection is the flip side of the cross event.

If the divine experience of history, from the sending of the Son up until the cross, has made an eternal impression from the temporal order to the Trinitarian life, how much more of an impression must the divine experience of history, from resurrection to the eschatological consummation, have made in glorification? The joy of responsive love of the redeemed creation in the Son eternally affects the life of the Trinity. Moltmann says that this is why Christian doxology always ends with the eschatological prospect, looking for the perfecting of the kingdom.[88]

It is self-evident that the resurrection should carry more theological weight than the cross in a Christian eschatological outlook and that Christian liturgy is and should be reflective of the creation's future glorification. The perfected and mature creation, freely responding to the overwhelming redemptive love of God is, and should be, an indispensable part of Christian doxology. This doxological nature of the church's eschatological outlook on the future of the cosmos, based on the interpretation of Christ's resurrection as the anticipation and first-fruit of cosmic redemption, needs more attention in relation to Christian worship, especially in this grim time of global suffering and looming ecological crisis. The liturgical aspect of eschatological-christological Trinitarianism will be addressed in chapters 4 and 5.

With the eschatological perspective on the cross and resurrection of Christ, Moltmann achieves a highly eschatological doctrine of the

---

87. *Trinity and the Kingdom*, 160.
88. Ibid.

Trinity firmly based on Christ's history; this broader Trinitarian narrative forms the context within which the church participates in the present struggle for the liberation of history and nature, and thus, participates in the suffering and joy of the Trinitarian God.

CONCLUSION

In this chapter, I presented Moltmann's eschatological Christology in relation to the Trinitarian history of God. Moltmann's Trinitarian understanding of the Christ event opens up an eschatological outlook of the creation; in addition, the eschatological-Trinitarian perspective makes his Christology an open one, by looking at Jesus Christ as a person in the making, a person on His way to the fullest expression of Sonship. The inseparable inter-relationship between Moltmann's Christology and his doctrine of the Trinity demands a close re-examination of the significance of the Christ event in light of the eschatological-Trinitarian outlook of the creation. In this re-examination, Moltmann arrives at the conclusion that the resurrection of Christ is the epistemological foundation for cosmic hope while the cross is the ontological foundation for ecological redemption.

Because the cross event is seen in the framework of a Trinitarian history of God, it is perceived as the division and unifying event in the Trinity. In this divine event, the whole cosmos is involved. Christ is seen to participate in the death of all finite living things, which again springs from and forms part of the physical and natural world. Through the unity that the Holy Spirit brings to the division between the Father and the Son during His experience on the cross, the whole Trinity participates in the cosmos with its sufferings and groaning, embracing the totality of the created reality in the inconceivably deep and wide space created *in* the Trinity. In this splitting and uniting Trinitarian event of the cross, all of time and space—even the deepest and most tragic dimension of the creation—is engulfed by the Trinitarian relationship.

The cosmic-sized space that is created in the Trinitarian God by the cross event is deep and wide enough to contain all dimensions, however tragic and forsaken. Furthermore, from this moment on, this space begins to be permeated by the Trinity. This spatial thinking opens up an ecological dimension in thinking of *missio Dei* and enables us to imagine the extent and degree of the transformative redemption that will be revealed through the resurrection of Christ and manifestation of the Holy Spirit, that is, a cosmic perichoretic indwelling of the Trinitarian God.

The resurrection on the other hand is seen as the epistemological event through which we can perceive the beginning of the eschatological transformation of the cosmos. It also convinces us that the cross was an ontological foundation of the cosmic redemption. In this context, the resurrected body of Christ takes on a special ecological significance because it is the inaugural focal point, an embodied divine promise of eschatological transition and transformation that definitely includes the physical and natural realms.

Another important point in relation to Christ's resurrection is that it is a transition rather than a radical discontinuity from the old creation. The pneumatological interpretation of the resurrection of Christ as a rebirth in the life-giving Spirit gains a special relevancy in this regard[89] because it can indicate both the continuity and discontinuity, the faithfulness of the Creator and the radical newness the creation is reshaped into. This eschatological- Trinitarian interpretation of the resurrection of Christ also sheds light on the eschatological task of the Holy Spirit who is sent by the risen Christ, as an eschatological anticipation of cosmic redemption—the Holy Spirit is the agent of the new creation, the agent of the resurrection of the cosmos, the guarantee and pledge of glory in the future (Eph 1:13–14).

I presented christological and pneumatological themes such as the cross and resurrection, against an eschatological horizon. The eschatological task of the Holy Spirit invites a doxology that far exceeds the boundaries of the church's and humanity's concern and expresses the cosmic dimension of Trinitarian redemption. Christian worship involves the fore-seeing and fore-taste of the new creation in Christ through the Holy Spirit and thus participates in the suffering and joy of the Trinitarian mission process. In the following chapter, I will articulate the meaning and purpose of the church and its worship in light of the eschatological-Trinitarian interpretation of the Christ event.

---

89. See the section titled "Creation in the Feminine Imagery of Conception and Birthing" in chapter 2.

4

Christian Worship as Anticipatory Celebration and Missional Sending

IN THIS CHAPTER, I will present Moltmann's understanding of Christian worship in relation to his eschatological panentheism. I will first discuss the grounding of the church and worship in the Trinitarian history of God. Then, I will discuss the eschatological and, thus, missional nature of the church and worship, in the horizon of God's movement toward its cosmic redemption. In this understanding, the church will emerge as a messianic community that anticipates the eschatological reality through the Holy Spirit in the here and now. By the same token, worship also takes on an eschatological nature and missional dimension. As the focal point of the Christian faith community, worship finds itself in an irreplaceable position in the life of the church because, in and through it, the church receives a foretaste of this eschatological reality and acquires the energy to live the newly sampled coming reality out in the world. Thus, worship stands at the centre of gathering and sending, celebration and mission, grace and discipleship. Later, I will pay special attention to Moltmann's understanding of baptism and the Eucharist, the two sacraments in most Protestant churches, to help demonstrate the ecological potential in the liturgy when it is properly understood and practiced.

THE CHURCH AND WORSHIP IN THE TRINITARIAN HISTORY OF GOD

The Trinity and its History as the Foundation of the Church

So far I have presented God's relationship to the world as a Trinitarian movement toward an eschatological panentheistic embrace of the creation. While the beginning of the Trinitarian movement is the birthing of creation in the Spirit and its goal is the eschatological Sabbath of the creation, the apex of that divine-cosmic journey is the cross / resurrection event to which the church is constantly drawn to look retrospectively (to creation) and prospectively (to the *Eschaton*). In this Trinitarian perspective, the creation, as well as God, is on a common journey. Nothing is static, nothing is eternally fixed. Christ Himself is on His way toward His future for which the church is waiting. If that is the case, we should also see the church in the same eschatological perspective as well as in an historical perspective. Viewed in this eschatological / historical perspective, the church betrays its unique position in the Trinitarian history of God—too late for present creation and too early for full consummation.

Viewing the church in light of the eschatological panentheistic Sabbath changes the traditional and strong distinction between the church and the world and re-arranges them in the larger context of God's eschatological cosmic movement.[1] Traditionally, the church understood itself in a world of its own, defining Christian ministry and life exclusively in terms of preaching, public worship, the pastorate, and charity. Those outside the church were perceived as either a hostile power or prospects to be won.[2] I would like to argue, however, that, just as nature and grace complement each other in preparation for the coming glory instead of being perpetually oppositional counterparts, the church should see itself as an eschatological reality, with an eschatological task of becoming a sign and tool of God's coming Reign for the world and in the world. The church's failure to define itself in that way has resulted

---

1. This thinking started largely after the World War II in the Dutch "theology of the apostolate." For a detailed discussion of the change of thinking, in both Protestantism and Catholicism, concerning the relationship between church and world, see Bosch, *Transforming Mission*, 376–78.

2. Part of the blame for this dualistic view on the relation between church and world comes from a one-sided, negative appropriation of the word "cosmos" in the Gospel of John, e.g., 17:14, 16. The same Gospel, however, declares that "God so loved the world that he gave his one and only Son" (3:16).

in confusion about its status in relation to the world, as in the triumphalism seen in the identification of the church with the realised Reign, or in a tragic resignation, as in the church's self-portrayal of the ark of salvation from the dark troubled water of the hopeless world.

As can be expected, Moltmann's ecclesiology is eschatological. For him, the church is an eschatological church, with a special mission in relation to the coming Reign of God. More concretely, in his eschatological ecclesiology, Moltmann places the church in the context of the Trinitarian history of God by strongly binding his ecclesiology to eschatological Christology and eschatological pneumatology. In fact, the church owes its very existence to the Holy Spirit, who moves toward the eschatological glorification of the Son and the Father. The messianic community bearing the name of Christ is thus an eschatological community. It anticipates, embodies, participates in, and witnesses to God's eschatological Reign.

In line with the Trinitarian history as the context of his ecclesiology, Moltmann consciously expands his theological horizon to include the whole history of the cosmos—the totality of created space and time—beyond humankind and its history. Against this broader horizon, the church is also seen as bearing more than the hope and sign of human salvation.[3] Because the church is viewed in the context of God's history of cosmic redemption of the old creation by way of the new creation through the creative mission of the Son and the Spirit, the church should be seen as living in the tension between hope and experience, and in so doing it understands itself as part of this history of the creative Spirit.[4] Between the hope of the eschatological panentheistic Sabbath and the present reality of disappointments, pain, and groaning, the church lives out the history of the Spirit. In this way, Moltmann grounds his ecclesiology on the Trinitarian history of God that thrusts forward toward the cosmic redemption.

Along with the Trinitarian history of God, another grounding of the church is the Trinity itself. If human beings are created in the image of God, if the whole creation is journeying to become the dwelling place of the triune God, and if the church is called out of the creation ahead

---

3. Unfortunately, this point is not as clear as desired in Moltmann up to *Church in the Power of the Spirit*.

4. *Church in the Power of the Spirit*, 3.

of time as an anticipatory community that embodies a new creation, the church should be in the image of the triune God.

At this point, the ecclesiology of Miroslav Volf—a student of Moltmann—proves very helpful. Volf presents the church's correspondence to the Trinity as an object of hope and goal for human beings. He bases this claim on Jesus' high-priestly prayer that His disciples might become one "as you, Father, are in me and I am in you, may they also be in us" (John 17:21). This presupposes our communion with the triune God and aims at its eschatological consummation.[5] The correspondence between the fellowship of the church and the fellowship of the Trinity is soteriologically grounded, since faith incorporates the believer simultaneously into Trinitarian and ecclesial communion. If the church is already participating through faith and baptism in the Trinity, then "the relations between the many in the church must reflect the mutual *love* of the divine Persons."[6] I view Volf's grounding of the church in the Trinity as a natural progression from Moltmann's *The Church in the Power of the Spirit* to a more fully argued grounding of the church based on Moltmann's concept of the perichoretic social Trinity. While Moltmann focuses on the eschatological task of the church in the context of the Trinitarian movement, Volf emphasises the church's anticipatory capturing of Trinitarian fellowship although Volf does not forget that it is the hope and goal of the church.

I view this two-layered grounding of the church – that is, the grounding in the Trinity (as Volf does in *After Our Likeness*) and in the Trinitarian history of God (as Moltmann does in *The Church in the Power of the Spirit*) – not as two but one. The Trinity exists, as far as the creation is concerned, in the Trinitarian movement toward the consummation of the creation, which is its participation in the Trinitarian fellowship at the *Eschaton*. As the Trinity intertwines with creation and moves forward, the church is also seen in this eschatological Trinitarian movement and, at the same time, anticipates and aspires to perichoretic love found in the Trinity. This sojourning character of an eschatological church betrays the limits of analogy between the fellowship in the church and that in the Trinitarian God.[7]

---

5. Volf, *After Our Likeness*, 195.

6. Ibid.

7. Volf puts it in this way: "The correspondence of ecclesial to Trinitarian communion is always lived on the path between baptism, which places human beings into

In relation to our present discussion of ecology, the importance of grounding the church in the Trinity and its history lies in the expansion of the church's purview to include the cosmic dimension. Because the history of the Spirit, of which the church is a part, embraces both human history and natural history, the church's purview should also be expanded.[8] This is the horizon against which the church should locate itself, against which the meaning of worship should be identified, and against which the mission of the church should be defined. This all-encompassing horizon offers the backdrop and inspiration of the church's doxology and mission.[9] The eschatological consummation that the church envisions and the perichoretic fellowship for which the church yearns, thus, should also include a cosmic dimension.

Second, because the church as a faith community anticipates the future of the creation—becoming the dwelling place of the Trinitarian God—by being included into the fellowship of the Trinity, now a new possibility is opened up: the possibility for the celebration of *our* perichoretic fellowship, within the church, with the creation. Although that fellowship must be in a limited and broken way due to the penultimate and sojourning nature of the ecclesial fellowship as Volf maintains, still it aims at growing toward the fullness of the Trinitarian fellowship. When this possibility is executed in an appropriate liturgical practice, it could lead to an ecological awareness amongst the worshipping community, an awareness that celebrates our connectedness and interdependence and the gloriously multiple forms and relations in the whole creation.[10]

---

communion with the triune God, and the eschatological new creation in which this communion is completed. Here the correspondence acquires an inner dynamic, moving between the historical minimum and the eschatological maximum." Ibid., 199.

8. Moltmann argues that the history of the creative Spirit should be understood "in dialectical-materialist terms as 'the movement,' 'the urge,' 'the spirit of life,' 'the tension,' and 'the torment,'" which are experienced not only by the church or humankind but also by the whole creation. *Church in the Power of the Spirit*, 34.

9. Moltmann's ecological theology thus expands the traditional boundaries of theology to include nature. Unfortunately, his earlier ecclesiology, defined in terms of an anticipatory participation of the church in God's eschatological consummation of *history*, is not reworked to include the cosmic dimension.

10. This is why the *worship* becomes so important for ecology. This point will be dealt in more detail in following sections.

Anticipation as a Crucial Ecclesial Concept

In understanding Moltmann's ecclesiology and especially the eschatological function of the church in the history of the Trinitarian God, the concept of *anticipation* is crucial. According to Moltmann, the eschatological future of history does not merely lie in an unreachable future but is brought into the present by the presence and power of the Spirit. Because of this Trinitarian movement that enters into and transforms the present world, although in a disputed and hidden way, the present history should be understood eschatologically. Since "History and eschatology cannot be metaphysically divided, as this world and the next, in the world and out of the world,"[11] we need a mediatory category that links eschatology and the present history, doing justice to both the "not yet" and the "already" sides of those realities created by the eschatological Holy Spirit.

Anticipation is the concept that Moltmann uses to represent the mediating category between the eschatology and the present history.[12] Moltmann claims, "Anticipation is *not yet* a fulfilment. But it is *already* the presence of the future in the condition of history. It is a fragment of the coming whole. It is a payment made *in advance* of complete fulfilment and *part-possession* of what is still to come."[13]

Moltmann uses this concept of anticipation as a double-edged sword: to defend against the church's arrogant triumphalism that identifies itself as the realised Reign of God on the one hand, and on the other to overcome a tragic resignation and inward withdrawal of the church from the struggle of the present. As long as anticipation is not merely an expectation of a distant future but a real part of what is to come, it guarantees its full realisation in the future; at the same time, as long as

---

11. *Church in the Power of the Spirit*, 192.

12. While common English usage of the word "anticipation" suggests expectation, hope, and foreknowledge, Moltmann uses the word in a somewhat peculiar way to mean "reality or realisation of something in advance." This usage can be justified etymologically because its Latin root, *anticipatio* literally means the taking of something in advance, which resonates well with the biblical concept of first fruit and pledge. When Moltmann uses anticipation in an eschatological context, it should be noted that he does not mean a vague idea or mere foreshadow of the future but, a *pars pro toto*, a real part of what is to come in full dimension, a reality and a real presence of the future brought about by the eschatological Holy Spirit.

13. Ibid., 193. Emphases are mine.

it is a reality in the present history, it is a reality only partially realised and foretasted.

Worship in Eschatological Ecclesiology

When we apply the above-described pneumatological concept of anticipation to ecclesiology, the existence of the church itself, in the power of the Spirit, is an anticipation of the coming Reign of God in history.[14] By extension, however, worship—the centre of the faith community—also becomes an anticipation of the coming Reign of God. Therefore, worship—just like the church—takes on an eschatological nature with the tension between "already" and "not yet."

The way worship becomes an eschatological anticipation in the Spirit involves the perception and reception, in faith, of God's Trinitarian dealings with the world by worshippers. Christian worship receives its content from the Trinitarian history of God, which the faith community perceives through a christological lens. In worship, through the stories of Christ's life, passion, and resurrection, the church ritually relives the life, suffering, and final consummation of the whole creation.[15] Moltmann puts it this way:

> The assembled community perceives anew the complete history of Christ, his giving himself up to death for the salvation of creation, this glorification in the life of God for creation's future. The messianic feast renews the remembrance of Christ and awakens hope for his kingdom.[16]

That is why worship—made possible by the Spirit, who is sent through Christ after His eschatological resurrection—is already a sign and demonstration of the coming Reign. In this respect, worship should be seen as a pneumatological event that transcends the linear timeline to which we are so accustomed.[17] Just as the cross and resurrection is a Trinitarian event that encompasses the whole cosmos, worship in the

---

14. Ibid., 196.

15. For the cosmic interpretation of the Christ event in the context of the Trinitarian history of God, see the section titled "Eschatological-Cosmic Perspective on the Cross and Resurrection in the Trinitarian History of God" in chapter 3.

16. Ibid.

17. A similar perspective on Moltmann's view of time—more specifically in the Eucharist and Sabbath—can be found in Nicholas John Ansell, "The Annihilation of Hell," 148–51.

Spirit made possible by the Spirit, ruptures the ordinary bounds of space and time and brings in an eschatological reality into the present gathering of believers so that the church can experience, be immersed, and live the eschatological feast and Sabbath in *anticipation*.

The implications of this eschatological anticipation in worship might be expressed in two ways. First, because worship captures, in a partial way of course, the whole Trinitarian mission and its completion, it becomes a real basis of the church's doxology not only of the economic Trinity but also of the eschatologically immanent Trinity.[18] This reinforces Volf's claim that the church should reflect the immanent Trinity.[19] Second, worship becomes the church's sensory basis of its hoped for future. Through smiles, laughter, tears, smell and taste, by hand-shakes, hugs, singing and hearing of hymns, through the hearing of the Word, and by feeling the peace among the accepted and reconciled, the worshipping faith community in its spiritual and physical dimensions "see" the risen Lord in the Spirit. Then, as the disciples receive the mission commandment from the risen Christ, worship becomes the place where faith community is charged with cosmos-encompassing mission.

The Tension in Worship as an Eschatological Anticipation

Eschatological panentheism or the eschatological perichoretic Sabbath, for Moltmann, becomes accessible through worship by the power of the Holy Spirit, although always as an anticipated feast, and thus, a partial participation. For him, worship is a messianic feast that anticipates the fellowship of God's Reign; the freedom in the presence of the triune God is experienced and celebrated *in advance* in worship.[20] This anticipatory participation and taste of the eschatological bliss through worship makes hope real and fervent. This is the "already" side of worship.

Worship, however, is not just an ecstatic experience of what is to come. Worshippers come from the present reality, with all its violence, tragedy, and sighs of the ailing creation. These feelings and experiences are not, and should not be, blocked at the entrance point of worship. Rather, those realities form an integral part in the structure of Christian

---

18. For the concept of eschatologically immanent Trinity, see the section titled "Trinity in Glorification" in chapter 2.

19. See the section titled "The Trinity and its History as the Foundation of the Church" in this chapter.

20. Ibid., 261.

worship because of its christological formation. Christians can enter the worship of God because Christ reached out to them, and identified with them in their unworthiness, in their unrighteousness, in their godlessness, and in their suffering in the darkness, and most of all, because He died a death of *God*-forsakenness.[21] This is the foundation not only of the Christian doctrine of justification by faith but also of Christian worship. In union with Christ in His passion and resurrection, Christians can participate in the Trinitarian community of God in worship. Moltmann believes "this is what is meant when Christian meetings and services are begun in Christ's name and in the name of the triune God."[22] If this is true, Christian worship can and should always be mindful of the death cry of Christ in His solidarity with the suffering creation. This is why Christian worship cannot just be ecstatic and glibly hopeful, but should be mindful of, and expressive of, the present situation of the world and the resulting earthly experience of pain and suffering.

In addition to the dire and miserable experience and memory worshippers bring to worship, there is another dimension on the "not yet" side. Exactly because worship is an anticipatory experience of the freedom, the peace, the love, the *perichoretic* oneness in the messianic era, it acutely contrasts the experience Christians have in the world: the pain, oppression, hatred, fractures and disconnection, the loneliness, and the sighs of the whole creation. In this sense too, worship contains the "not yet" side of the coin. In fact, worship highlights the inhumanity of human society, with all its brokenness, as well as its exploitation of, and aggression on nature. This contrasts with the common perception that worship soothes and relieves worshippers of painful memories and psychological burdens; instead, as an eschatological anticipation, worship makes the Christian experience an even more acute pain, as well as restlessness, strife, and homelessness in this unredeemed state of the world order. This is the "not yet" side of worship.

Worship and Mission

If worship is in an eschatological tension between the present reality and the eschatological reality, between remembrance and hope, of "already" and "not yet," then this tension is not a trivial emotional by-product. Rather the tension should be seen as an essential experience of worship,

21. *Crucified God*, 145.
22. *Church in the Power of the Spirit*, 261.

coming from the very anticipatory and Eucharistic nature and structure of Christian worship.[23] The tension in worship as an eschatological anticipation, then, should be seen as a work of the Holy Spirit that remoulds and inspires the congregation. This is the point where grace calls on discipleship, where God's initiative love asks for a response, where God's passion in *missio Dei* spills over into the life of the church. In this sense, worship is the smelting furnace in God's workshop in which worshippers are transformed into the image of God whom they worship.[24]

The Holy Spirit awakens the dormant missionary in every Christian by creating and using the tension that arises from the foretaste of the coming Reign and the bitter contrast of the present state of the world, against which both God and the worshippers suffer and protest. Out of the tension in worship springs the motivation and energy for the *mission* to protest and transform the world. Exactly here, I would argue, lies the significance of the eschatological panentheism and its anticipatory celebration in Christian worship.

This awakening, this charge of motivation and energy, which in fact is a reflection of the missional God, can set the everyday life of

---

23. If worship involves perceiving and appropriating the Trinitarian history of God through the story of Christ, and if the resurrection of Christ is already the anticipation of this eschatological feast and Sabbath, what happens in worship in the name of Christ is *remembering* the *future*. More simply put, worship in the Christian church is a christological remembrance of an eschatological future in the Holy Spirit. This is what I mean by the Eucharistic nature of worship. Moltmann's view of the Eucharist as remembrance and hope will be discussed in more detail below.

24. Reformed theology expresses a strong objection to the understanding of worship as human work toward God and rightly emphasises that worship is God's work. If worship is God's work, however, it is only natural to expect a change in worshippers as a result of encountering and appropriating God's initiative work in the past, present and future. And the goal of God's work on the congregation is mainly understood as transforming the worshipping congregation into the image of God. Romans 12:1–2 also links worship and the transformation of believers, although here Paul refers to living sacrifices, which correspond to mission in the frame of this thesis. Here, the dichotomy whether worship is about us or about the One worshipped loses its force. Worship should be understood as an interaction between God and the worshipping faith community, between initiative grace and thankful and faithful response. In and through that interaction, the worshipping community in the Holy Spirit experiences, enjoys, and aspires to take on the Trinitarian God. For that reason, with Old, I would like to call worship God's workshop. Old, *Worship*, 6. Zwingli's idea of a congregation's transubstantiation is very sympathetic to the idea of worship as God's workshop although he was referring specifically to the Eucharist. Zwinglian idea of transubstantiation will be discussed in more detail below.

Christians in the context of the Trinitarian history. In fact, when set in this context, the whole of Christian life begins to look drastically different.[25] It is no longer a meaningless suffering or toiling under a fatefully fixed system but a *mustard seed* and *yeast* that anticipates the coming Reign. Christians who understand and live their lives in this way no longer perceive the present as an eternal order under which they have to endure in obedience. They no longer despair of, nor acquiesce to, the dire worldly condition. Rather, through worship, they recognise the great arc spanning the remembrance (the life and passion of Christ) and the hope (the resurrection of Christ and the ensuing hope for creation) and situate their lives, their current hope and struggle with the powers that be, in this arc.[26]

When a Christian begins to see his or her life in this way, that life begins to take on a missional nature because it participates in the Trinitarian missional and redemptive movement. By bringing anticipation, resistance, representation and self-giving to their everyday lives, Christians become eschatological agents in God's salvation history, to transform the creation into the hoped-for new creation.[27] In that sense, seemingly ordinary Christian life outside worship obtains even a salvific dimension, although we have to admit it is in a hidden and disputed way.

Therefore, every time someone is baptised, the gathered congregation awakens to the renewed hope of the new heaven and new earth. Every time they celebrate the Eucharist, they foretaste the final *shalom* that will be realised throughout the universe. Every Sunday, when Christians leave their place of worship, they clothe themselves anew with a burning sense of justice and an unquenchable thirst for peace and compassion for everything in the world, knowing and waiting painfully for the arrival of the healed, matured, renewed, and God-indwelling new creation. In this way, Christian worship, through the pain and hope it enthuses into Christians, could and should function as an energy for ecological mission and a thirst for ecological bliss for our home, Earth.

25. Richard Kearney articulated the significance of stories that frame and thus make human life sensible. He argues that identity is formed and re-formed through stories and the re-telling of stories. Moreover, he argues existence "is not fully narrative until it is re-created in terms of a formal verbal recounting." Kearney, *On Stories*, 79–81, 132. I would argue that, for Christians, this defining or re-defining story is, and should be, the story of Christ and the Trinitarian missional God.

26. *Church in the Power of the Spirit*, 261.

27. Ibid., 194–95.

Worship / Mission and Ecology

So far I have described how worship can be a source of missional motivation and energy. Now we must question how exactly this worship and missional energy, flowing out of it, can work for an *ecological* mission. In other words, we must ask how exactly worship can help and encourage worshippers to become ecological agents, leading them to a better appreciation of nature in its own value and beauty, and to transform their individual and communal life toward a reconciliation with nature.

In this section, I would like to present the importance of *moral formation* and *moral imagination* in the faith community, and then the bodily and communal dimensions of worship as the link to ecological awareness. Finally, as a conclusion, I will present the faith community at worship as a model and an outpost that eschatologically anticipates ecological redemption.

Worship as the Locus of Moral Formation and Moral Imagination

The biggest challenge that faces the ecological movement is neither scientific nor technological. Nor is it a matter of education, spreading information, and raising awareness, although all of these are important. Mary Evelyn Tucker expressed this concern:

> It is becoming increasingly evident that abundant scientific knowledge of the crisis is available and numerous political and economic statements have been formulated. Yet we seem to lack the political, economic, and scientific leadership to make necessary changes. Moreover, what is still lacking is the religious engagement to transform the environmental crisis from an issue on paper to one of effective policy, from rhetoric in print to realism in action.[28]

The reality is that we as the dominant species on earth do not have sufficient will to act, or to change our current destructive path. This is so despite the mounting evidence concerning the current state of the crisis as well as a multitude of suggested agendas for both individual and collective agents such as families, communities, governments, and international organisations. Ernst Conradie points to the unwillingness of the United States to agree to the moderate measures of the Kyoto Protocol as evidence that information itself is never enough. Where the desire of

---

28. Hessel and Ruether, *Christianity and Ecology*, xix.

the affluent is to maintain—and even expand—their lifestyle while the poor aspire to catching up with the lifestyles of the affluent, the key to ecological transformation does not lie in the matter of *information* but in moral *formation*.[29]

But how can a faith community muster enough will to act decisively amongst fellow Christians and beyond? In this connection, worship can make a key contribution. In relation to an ecological re-envisioning, Santmire deplores the neglect of worship in Rasmussen's *Moral Fragments and Moral Community*, and other like-minded pragmatic, bottom-line Americans.[30] Here, I am of the same opinion as Santmire who argues that while Birch and Rasmussen's processes to form moral formation are extensive, they miss the point by discursively enumerating many means of moral formation and delegating worship, probably unintentionally, to one of them. Worship—the centre of the church—is a constant in the history of the church, the pre-eminent place for Christian theology, an integral and principal part of the church's mission.[31] In this sense, worship is not simply what the church does on Sunday. Rather, through worship, the church becomes what it is supposed to be—the smelting furnace in God's workshop, where worshippers are transformed into the image of God whom they worship, and made aware of the eschatological reality to which we and the whole cosmic community aspire. To employ Santmire's terms, worship is the church's "mode of identity-formation" and a "generative matrix," through which all the activities are sustained, such as the communal inculcation of values, the proclamation of social justice and good stewardship, and the kerygmatic call to evangelisation.[32]

Worship is, and should be an event through which we stand before and in the Triune God and its history, and by which we become transformed. By way of anticipation in the power of the Holy Spirit, Christians

---

29. Conradie, *Church and Climate Change*, 62–63.

30. "This kind of relegation of worship to the end, intended or not, must be corrected in the context of our discussion here, emphatically. Because, it appears, anthropocentric traditions in worship in the Christian West have left the Christian community ill-equipped to be the kind of exemplary moral community that Rasmussen so appropriately calls it to be. In fact, it could be argued that one of the main reasons that the church has found itself so ill-equipped in so many ways to respond to the environmental—and other—dimensions of today's global crisis is precisely those anthropocentric traditions of worship." Santmire, *Nature Reborn*, 75.

31. Fensham, *Emerging from the Dark Age Ahead*, 161.

32. Santmire, *Nature Reborn*, 76.

at worship are enlightened, inspired, and immersed into an alternative vision and reality, although in an ephemeral and partial way.[33] During and through this anticipatory glance and foretaste of redemptive future, we see the vision of the whole creation—healed and Spirit-filled—shouting the praise of its Creator; we dare to dream of a cosmic consummation, in which the creation becomes the home of God and God will be all in all; the future of Christ's glory in God's glory is enabled, for which we wait with the whole creation. It is out of this glance, this foretaste, and this immersion into redemptive future that the *moral will* to march toward this blissful vision is formed and the *moral imagination* to reject the deceptively ordinary, yet destructive way of life is ignited.

Of course, this moral formation and ignition of moral imagination through worship do not always happen automatically. When worship is formed in an anti-ecological, a-cosmic, anthropocentric "Great Chain of Being" and Gothic spirit;[34] when worship degenerates to a meaningless repetition of conventional rite without realising the breadth and depth of the promise on which worship is based and into which worship aspires to enter; when worship is marketed as religious entertainment, fashioned to satisfy the palate of contemporary religious consumers deeply tuned to the current world-order; when the congregation turns a deaf ear to the painful groaning and the whispering invitation to hope of the Spirit, then the eschatological significance of worship becomes lost. It becomes a religious legitimisation of the current way of life. That is why our understanding of worship and how we engage ourselves in worship is so important. When worship is seen merely as a human work to ensure divine favour for whatever design congregants are already pursuing, worship becomes the pinnacle of human self-centeredness and selfishness that exponentially exacerbates the languishing of creation. As the Reformers have so emphatically argued, however, the purpose of worship is not to brace our current way of life with an even higher power, but is designed to open us up to God's gracious salvific act and let that initiative love change us.[35] However, if we refuse to succumb to the pressure of today's society to compartmentalise our religion, to try to be

---

33. In this regard, the confession of Peter on the transfiguration mountain (Mark 9:5) still resonates.

34. See chapter 1.

35. Wolterstorff, "Genius of Reformed Liturgy" Online: http://www.reformedworship.org/magazine/article.cfm?article_id=24; internet.

palatable to religious consumers, and if we vow to base our worship once again on a sound theological understanding and eschatologically oriented liturgy, then, I would argue, worship can awaken the congregation to its eschatological significance and messianic mission. In so doing, it can become a true epicentre of transformation, a true smelting furnace to form a moral will, a true watchtower to envision an alternative way.

Worship in Faith Community as a Bodily and Communal Experience

Even if we agree that worship is the locus of moral formation and moral imagination, however, we have not yet articulated exactly what aspects of worship lead to an *ecological* moral formation. I argue that the bodily and communal dimensions of worship link worship to an ecological moral formation.

Worship as an anticipatory experience of the eschatological Sabbath, in which all creation participates in the fellowship of the Trinity, is experienced not through an individual soul's union with the Divine but through a bodily and communal experience. The word "experience" itself suggests an involvement with something other than the solitary individual self, going-out-of-oneself to encounter and connect to something outside. This encounter or connection occurs essentially through a *bodily* function to reach a *communal* experience. I would like to submit that this bodily and communal understanding of worship is where ecological theology of worship and a Gothic-spirited theology of worship diverge because the former emphasizes the bodily and communal dimensions while the latter focuses on a disembodied experience of the individual soul.

It is important to note here that these bodily and communal aspects of worship derive from the fact that eschatological worship is an anticipatory participation in the perichoretic community of Trinity. If worship is the smelting furnace of the Holy Spirit that moulds the worshipping community into the image of the perichoretic divine community, it is natural that the worshipping community begins to perceive itself as a brotherly and sisterly community and experience the communal pleasure of peace, joy, freedom, acceptance and being accepted, forgiving and being forgiven, and a realisation of inter-dependence. Additionally, these experiences come through, and lead to, bodily and sensory mediation, such as greeting, touching, hugging, smiling, singing, listening, the seeing and feeling of sprinkled or poured water in baptism, and the eating and drinking of the Eucharistic elements. It is the Christian body, individual and communal, that comes to the forefront in a worshipping experience and participates in, absorbs and reacts to the communal feast.

In this experience, the strict separation of subject and object fades away and the strong individualistic "subject" awareness, which is a trademark of modernity, gives way to the community awareness that postulates the fundamental interconnectedness of everyone and everything and takes for granted the co-establishment of identity and relationship. When this community awareness in worship is thought, felt, interpreted and pursued on the dimension of human history, it becomes the anticipation of human liberation. When it is deliberately expanded to include the whole creation by means of a theologically-informed liturgy and theological reflection, however, it becomes the anticipation of cosmic redemption, an anticipatory communal experience of eschatological Sabbath, a cosmic *shalom* in which the whole of creation enjoys the participation in the fellowship of the Trinity.

Of course, the anticipation of human liberation and cosmic redemption are not mutually exclusive. Rather, they are concentrically related. Cosmic redemption contains human liberation and serves as its context, while human liberation is the sign of the cosmic redemption and serves as a precursory catalyst. The worshipping community that anticipates the eschatological Sabbath in the Holy Spirit is in Christ. By being in Christ, the worshipping faith community also anticipates and betrays the cosmic *shalom* at the *Eschaton* that Christ anticipated in His resurrection. Therefore, the Christian community's awareness that comes from the experience of worship, which participates anticipatorily in the fellowship of the Trinity, is closely related to the Christian ecological awareness that feels kinship and interdependence with the whole creation and yearns for a peaceful inter-dwelling in the creation.

Is it not natural, then, that we should acknowledge that worship is an experience of and participation in the communal God, who created and sustains and remoulds the creation so that this created bio-physical reality can finally achieve the full participation into that divine fellowship? Is it also not right that we praise God because we experience God in these bodily and communal dimensions?

Ecological Mission as an Option for the Poor

In *Experiences in Theology*, Moltmann develops an expanded ecclesiology in dialogue with Ahn Byung-Mu's *minjung* theology. According to Ahn, "the crowd," "the many" or "the people," who in traditional theology are merely called "the nameless background for Jesus' ministry," a "background choir" for the appearances of the star, are actually the main reason for Jesus' coming and ministry. Moltmann reminds us of the relationship between Jesus and *ochlos*:

1. Jesus "teaches" the people. He does not teach the people the Torah. He teaches them the gospel of the kingdom, the messianic beatitudes and the discipleship ethic of the Sermon on the Mount.
2. It is "the nameless crowd" who are Jesus' people and family, not His own biological family, nor Israel, His cultural people either.
3. *Ochlos* are "God's people," the people of God's kingdom, the people of the poor who are 'called' and therefore chosen.
4. The healing miracles of God's kingdom and Jesus' table-fellowship with "sinners and tax collectors" take place among these people.
5. When He goes to Jerusalem, He calls these people to the discipleship of the cross.
6. He sheds His blood and gives His life for these people.[36]

Based on these observations, Ahn argues:

> Jesus is unconditionally on the side of the *ochlos*. He quite evidently has no intention of turning the *ochlos* into an anti-Roman fighting force, but He proclaims to them the kingdom of God as their future, the future which already belongs to them here and now, and He fills this people with new hope and with the vision of a way that can be their own. "Jesus fights together with the suffering *minjung* (*ochlos*) on the front of this Advent [of God's]."[37]

Although Moltmann provides some critique on Ahn's Christology, he sympathizes with Ahn's biblical analysis. Furthermore, building on the *minjung* theology's emphasis on *ochlos*, Moltmann proceeds to suggest an expanded ecclesiology—the *manifest* church and the *latent* church. The *manifest* church consists of believers and followers of Jesus. This church enjoys Jesus' active identification; real presence by virtue of

---

36. *Experiences in Theology*, 254.
37. Ibid., 255.

Christ's identification with the community of His people fills the proclamation, the sacraments, the fellowship and the *diakonia* of His congregation with authority. Moltmann cites two passages here: "He who hears you hears me" (Luke 10:16); and "As the Father has sent me, even so I send you" (John 20:21). On the other hand, the latent church consists of *ochlos*—the poor, hungry, thirsty, imprisoned, and the children. Moltmann cites again two passages: "As you did it to one of the least of these my brethren, you did it to me" (Matt. 25:40); and "Whoever receives one such child in my name receives me" (Matt. 18:5).[38] Moltmann argues that the whole Christ is present in the manifest community of believers, and in the latent community of the poor.

This expanded concept of the church, as Moltmann himself acknowledges, poses a tremendous tension in our ecclesiological thinking. In a sense, however, the church has always related to the poor through its teaching, especially by the book of James, and in its missional outreach.[39] I believe that the *minjung* theology's contribution in highlighting Jesus' special relationship with the *ochlos* in solidarity and representation is highly significant. Because Jesus identified with the *ochlos*, the church should always have a special relationship with the poor, oppressed, marginalized, outcast, weak, discriminated, and scapegoated—as liberation, feminist, gay / lesbian, black, and post-colonial theologies have taught us. Furthermore, I believe that the boundary of the church is unimportant and should be flexible, since the church is a tentative existence that serves as a sign and tool for the coming Reign. Nevertheless, I hesitate to sympathise with Moltmann's attempt to confer the name "church" to the *ochlos*. For me the church is always called to worship and thus to mission. If we apply this concept of the church, even with the qualifier *latent*, will it not result in emptying the meaning of the church? Where is the benefit of bestowing the title "church" to *ochlos* or *minjung,* who do not worship the Christ in the Spirit, and who do not appropriate the promise in Christ, and thus have no sense of mission?

Moltmann's suggestion, however, brings up some important aspects of the so-called manifest church's worship and mission. If Jesus identified with the *ochlos*– sinners, those excluded from Jewish society,

---

38. Ibid., 266–67.

39. See for example the Roman Emperor Julian's complaint that Christians not only take care of their own poor, but also the rest of the Empire's poor as well. *Julian, letter 22*, in *Works of Emperor Julian*, 3:71.

impoverished country people, landless people, or the lost—it begs the question, "Who are the *ochlos* today?" Should not the silently suffering and dying creatures on the edge of human civilisation, being forced to evacuate from their natural habitat and thus made homeless, the least in the creation, be included in the *ochlos* today? Should not the mercilessly assaulted creation itself be included in the *ochlos*? It is clear, then, that the church's program since liberation theology—options for the poor—needs to be expanded to include the present day suffering creation. Here again, the church's aspiration and mission toward human liberation is concentrically expanded to include the healing and well-being of the creation. This is another reason—in addition to the bodily nature of humans and the church—why the mission of the church for liberation should also be an ecological mission.

The Faith Community at Worship as the Epicentre of an Ecological Mission

Before moving on to an ecological understanding of baptism and Eucharist, I would like to summarize my discussion of church and worship. I have based the existence of the church on the Trinity and its history. As Charles Fensham persuasively argues, the reason for the church's deplorable search for, and indulgence in, domination, brute power and influence is its negligence of its foundation: the Trinitarian God, as a perichoretic self-giving missional divine community.[40] The real relevance of the church for today's world, therefore, comes from its relevance to God, not from its futile and unauthentic effort to be relevant to the contemporary culture of domination, mastery, and marketing.[41] When we base the existence of the church on the Trinitarian God, this Trinitarian foundation of the church compels us to expand also the church's missional horizon to include the cosmic dimension.

At the centre of these ecclesiological dimensions—the foundation of the existence and mission of the church—stands worship, as the smelting furnace through which the Holy Spirit transforms the worshipping community into the image of the divine missional perichoretic community. Furthermore, the intrinsic tension in an eschatological anticipatory worship evokes a moral formation and a moral imagination for the

---

40. Fensham, *Emerging from the Dark Age Ahead*, 146–49.
41. Ibid., 150.

church so that it can live out a missional life in a bold and creative way according to the eschatological vision experienced in worship, that is, a life of transforming mission. The bodily and communal dimensions of worship, with the understanding of their origin in the divine perichoretic Trinity that seeks an indwelling in the created reality, can serve as the link between the Christian community awareness and Christian ecological awareness. The worship experience, in its nature of eschatological anticipation, can serve as a strong source of ecological awareness, as well as an ecological moral formation and the source of moral imagination. It is appropriate then to call the worshipping faith community as the epicentre of ecological mission.

BAPTISM AS A SIGN OF COSMIC REBIRTH

Baptism, along with the Eucharist, is the most important liturgy in Christian worship. Therefore, in constructing an ecological worship, it is essential to understand the eschatological meaning and its cosmic breadth and depth of the promise it carries. Thus, now I turn to Moltmann's understanding of baptism in order to envision an ecological Christian worship.

Moltmann's Baptismal Theology

Under the overall understanding of the church and worship and its location and meaning in the Trinitarian history of God, Moltmann discusses the eschatological and cosmic significance of baptism. The most fundamental premise of Moltmann here is that baptism is a sign of the messianic era that places the believer, first, in the messianic community, and second, in the Trinitarian history of God by uniting the believer with Christ.

Moltmann develops the ecological significance of Christ's resurrection and the Pentecostal outpouring of the Holy Spirit mostly in *The Way of Jesus Christ* and *The Spirit of Life*. In *The Church in the Power of the Spirit*, however, his discussion of baptism, which is a sign of the death and rebirth to eschatological life, mainly relates to the historical liberation of humankind. His basic treatment of baptism in an eschatological and Trinitarian perspective, however, translates quite easily into ecological significance.

Christian Worship as Anticipatory Celebration and Missional Sending

In this section, I will survey Moltmann's criticism of baptismal theology in the German church from an eschatological and Trinitarian perspective. Some of his criticisms below are perhaps not so relevant outside his German context. Nevertheless, his discussion is important for us to note so that we can develop an ecologically appropriate baptismal theology. After presenting Moltmann's critical review and his own baptismal understanding, I will articulate its ecological significance on the basis of Moltmann's later development of Christology and pneumatology.

Present German Baptismal Practice and Theology Reviewed

Historical Approach to Baptism: A Founder Christology[42]

Employing an eschatological and Trinitarian perspective, Moltmann criticises contemporary Protestant baptismal practice and theology because they blur the fundamental nature of baptism with its historical approach and lack of eschatological perspective. In Protestant orthodox theology, baptism is one of the holy rites ordained by God, through which the saving grace of Christ is appropriated, making it an instrument of salvation. As a "holy rite," it is traditionally regarded as linking visible and invisible realities in a metaphorical way. Through baptism men and women are regarded as having been born again to eternal life. In that sense, traditional Protestant theology has held that baptism is "the sacrament of initiation and the door of grace."[43]

Furthermore, in this historical approach of traditional Protestant theology, baptism is validated by a founder Christology; that is, it gains legitimacy because it is instituted or founded, by Christ himself. Thus, as long as it is performed through the church and in the church in accordance with its institution, it becomes an efficacious sacrament. It is nothing more than Christ's words of institution and His promise uttered in the past that validates and activates the saving grace in baptism.[44] In

---

42. Moltmann's eschatological approach to the Christian tradition criticises the historical approach to many of the church's practices resulting in repetition of the so-called "apostolic form." But his eschatological approach does not exclude the historical perspective because he believes that the historical foundation of Christian tradition and the Scriptures are fundamentally eschatologically oriented and thus the historical approach employed in this perspective contributes to the right orientation of the church's praxis today. Lee, "An Eschatological Apostolic Understanding of the Authority of the Bible," 36–41.

43. *Church in the Power of the Spirit*, 226–28.

44. Ibid.

this perspective, there is little room for baptism to function as an eschatological sign of the coming Reign of God according to which the whole of our lives should be transformed.

A Sign of the Status Quo

Moltmann attributes the Reformers' decision in favor of the preservation of infant baptism—despite the open theological problem of establishing its origin in the New Testament—to the church's aspiration to maintain Christendom. Moltmann views infant baptism as the basic pillar of Christian society and the foundation of a national church since through infant baptism Christian society perpetuates itself.

In this form of baptism, the church fulfils a religious function, such as socialization and integration ordered by society: "Baptism at birth; confirmation at puberty; the wedding at the beginning of married life; extreme unction before death; and church burial afterwards."[45] When baptism is embedded in the social structure of the day, it is inevitable that "the form and interpretation of baptism takes on the impress of the society's predominating interests."[46] Baptism, therefore, loses its character as an eschatological sign, a sign of commitment and confession of resistance and hope in the change that an eschatological future brings into the present. Instead, the church that performs such social functions is a part of a huge mechanism of oppression, a sign of *status quo*. Thus, a "change of baptismal practice without a change in the public form and function of the church in society is not possible."[47]

Historical Review of Baptism and the History of Christ

Moltmann attempts to demonstrate the eschatological nature of the sign of baptism. For that purpose, he revisits the history of Jesus and His baptism as well as the primitive Christian church's understanding of baptism found in the New Testament.

---

45. Ibid., 231.

46. Ibid., 232.

47. Ibid., 229. Moltmann is here referring to the current German approach where the church is supported by the state through taxes and is thus integrated with the apparatus of the state. The limit of his criticism of baptism in the context of Germany is discussed in this chapter's "2.1.3. Moltmann's Understanding of Baptism as an Eschatological Sign Bound up with Christology and Pneumatology.".

Baptism of Jesus by John the Baptist

To clarify the Christian meaning of baptism, Moltmann first discusses John the Baptist's repentance movement and Jesus' baptism by John as the historical origin of Christian baptism. John the Baptist introduced a movement for repentance in Israel, proclaiming an imminent judgment of God and calling the people to repent of their unrighteousness. The fact that John baptized in the Jordan River is noteworthy, and eschatologically significant, representing a new exodus from bondage and the eschatological entry into the promised land of the divine kingdom. According to Moltmann, "It was the eschatological sign of the setting forth out of present oppression towards the immediately imminent freedom of the divine rule."[48]

Jesus' baptism by John the Baptist allows us to posit that Jesus, and later the Christian community, inherited some aspects of John's eschatological message. But at the same time, the fact that Jesus parted from John and began preaching His own message shows that His eschatological gospel differed from John's eschatology of judgment. Moltmann notes:

> Where John proclaimed the kingdom of God as judgment, with a view to repentance, Jesus evidently proclaimed the kingdom of God as the justice of grace and demonstrated it by acts of forgiving sins. For Jesus the gospel of the kingdom was an eschatological message of joy.[49]

In order to understand the Christian meaning of baptism, one must first understand the commonality—the imminent eschatological expectation, the provocation of the powerful and the collaborators, and the call to freedom it addressed to the people—and the divergence—Jesus' emphasis on the justice of God's grace and liberating good news of joy—between the messages of John the Baptist and Jesus. In conjunction with John's eschatology, one may comprehend Christian baptism as being a sign of the coming of God into a person's life and his/her turning to the common future.

---

48. Ibid., 233.
49. Ibid., 234.

Baptism in the Primitive Church

In the glow of Jesus' resurrection and in the experience of the Spirit, the primitive church began baptizing soon after Easter and consequently understood their baptism eschatologically as the pledge of what is to come and as the dawn of God's glory in a person's life.[50] Moltmann notes the correspondence between the baptism of the primitive church after Easter, and the earthly Jesus' turning from John's threat of judgment to the revelation of the gospel of the kingdom in the forgiveness of sins.[51]

In the primitive church, both in the Jewish and mixed communities, baptism is understood as a sign of new fellowship on the basis of the outpouring of the Spirit. In the Jewish Christian community, where they understood themselves as the genesis of the eschatologically renewed Israel, "Christian baptism counted as being the symbol of this messianic renewal of God's people." In the mixed Jewish and Gentile Christian communities or just Gentile Christians, where they conceived of themselves as the new people of God, "baptism became the emblem of the 'new creation' in Christ."[52] In this understanding, we need to note that baptism has an intrinsically communal dimension because it is a sign of the beginning of an eschatologically new people. Moltmann emphasises this communal dimension of baptism by pointing out the Pauline understanding of baptism in relation to charismatic community (1 Corinthians 12). Through baptism, people are integrated into their community, being entrusted with their particular ministry. In this sense, baptism is a call into a messianic community.[53]

Moltmann's Understanding of Baptism as an Eschatological Sign Bound up with Christology and Pneumatology

Moltmann notes the multiplicity of notions about baptism in the New Testament and, incorporating its diverse aspects, suggests we understand baptism in the framework of the whole history of Christ. In other words, he proposes to overcome the sole dependency on the founder Christology to validate baptism and to recover its eschatological significance beyond a sign of *status quo*. For that purpose, Moltmann proposes to see the baptismal event as an integral part of the history of the *living*

---

50. Ibid., 234–35.
51. Ibid., 235.
52. Ibid.
53. Ibid., 238.

Christ, who, in the Holy Spirit, was baptized, crucified, risen and is on His way to the consummation of His history, which is again bound up with the consummation of the whole creation.[54]

More concretely, when we consider Christ and His continuing eschatologically-oriented history as the framework of baptismal understanding, we incorporate diverse perspectives on baptism and their linking of baptism to various moments in Jesus' history. In Mark, baptism is related to the baptism of Jesus, which in turn points forward to Jesus' passion and death; in Matthew, baptism is related to the commission of the resurrected Christ; in Luke, baptism points to the pouring out of the Spirit at Pentecost; and in Paul, baptism is fellowship with the cross of Jesus and the promise of future fellowship with His resurrection. Furthermore, placing baptism in the framework of the whole of the eschatological history of Jesus helps us overcome a barren historical "founder Christology" perspective and move on to find the meaning of baptism in the context of pneumatic Christology and christological pneumatology, thus finally setting it in the Trinitarian history of God.[55]

In this perspective, baptism is seen both from the history of crucified Jesus and the eschatological history of living Christ pneumatically active here and now. Based both on the historical-theological life of Jesus Christ and on the eschatological opening to the coming Reign of God by way of an eschatological understanding of the coming and task of the Holy Spirit, "baptism points to the liberation of people which took place once and for all in the death of Christ" and, at the same time, "it reveals the crucified Lord's claim to new life and anticipates in man the future of God's universal glory."[56] Then, it not only leads us back to the cross of Jesus Christ but also to the power of the resurrection and new creation in the Holy Spirit.[57] In this way, baptism is an "eschatology put into practice" and is "Christian hope in action."[58] In a typically Moltmannian fashion, this historical-eschatological understanding of baptism helps us

---

54. Ibid., 238.

55. "The history of Christ and the history of the Holy Spirit are so interwoven that a pneumatic Christology leads with inner cogency to a christological pneumatology. We shall therefore have to see baptism, as well, as part of the Trinitarian relations of the eschatological history of God's dealings with the world." Ibid. 236.

56. Ibid., 240.

57. Ibid., 239.

58. Ibid., 235.

open up to the eschatologically transformative power of baptism for the here and now.

At this point, some critique of Moltmann's position on baptism might be in order. Although his discussion of baptism provides many helpful insights that we need to consider in order to develop an ecological baptismal theology and practice, we have the impression that some of his criticisms on Protestant baptismal practice and baptismal theology are mainly about the German national church and thus miss other important aspects of baptism in other parts of the world. For example, his discussion of baptism from the eschatological perspective helps us to overcome the narrow historical perspective and address the dynamic power in baptism available through the Holy Spirit for the here and now. His cynical designation of baptism as a sign of the *status quo*, however, is not fair to the baptisms ministered in places outside the German context. In places where Christians are in the minority, baptism is often the beginning of public or secret persecution, resulting in the separation of those baptized from their family and community. In those areas, then, baptism is hardly a sign of the *status quo* but an entrance into a new community that signals and witnesses the beginning of an alternative way of living. Moltmann's discussion of infant baptism poses another problem from a Reformed perspective. Moltmann pleads for adult baptism and against infant baptism on two accounts. First, theologically, there is meager historical basis in the New Testament for infant baptism, and the fact that only infants from Christian families are baptized invalidates the argument that infant baptism is a sign of prevenient grace. Also, from a political perspective, infant baptism forms the foundation of a national church, which Moltmann opposes. Although I agree that infant baptism is not amply supported by the New Testament, Moltmann does not pay enough attention to the prevenient grace argument in his theological discussion on infant baptism.[59] Regarding what Moltmann calls a political problem with infant baptism, his statement that "Anyone who affirms infant baptism, for whatever theological reason, thereby affirms at the same time this public form of the church [national church],

---

59. If baptism is a sign of an eschatologically new community that believers enter through the confession of faith (in the case of adult baptism) or through confession of the parents (in case of infant baptism), then not ministering to non-Christian infants can hardly be an argument against prevenient grace in infant baptism. Through the infant baptism and evangelism that follows, that baptism can indeed be a sign of prevenient grace.

or Christianity"[60] sounds strange and irrelevant both to Asian and Canadian ears. Finally, Moltmann's eschatological baptismal theology stops short of a fully ecological doctrine. By and large, it focuses on the liberation of humanity from oppression, as is true of the whole of Moltmann's early trilogy. His ecological concern comes to the forefront starting from *God in Creation* and consequently he develops the ecological significance of Christ's resurrection and the Pentecostal outpouring of the Holy Spirit in *The Way of Jesus Christ* and *The Spirit of Life*. Unfortunately, however, Moltmann does not revisit his earlier ecclesiology with this broader ecological perspective; therefore, ecclesiological concepts, including baptism, need a clearer articulation in relation to ecological concerns.

Baptism as a Sign of Cosmic Rebirth

Despite the above-mentioned shortcomings, formulating an ecological perspective of baptism will not be difficult if we marry Moltmann's eschatological and Trinitarian treatment of baptism with his ecological understanding of Christology and pneumatology. Let us put aside his socio-political argument regarding infant baptism for a while and focus on his basic christological and pneumatological contention that baptism is an eschatological sign.

If we agree with Moltmann's christological and pneumatological understanding of Christ's resurrection as Christ's anticipatory transition to an eschatological new life in the Holy Spirit and His resurrected body as an embodied promise of the eschatological deification of the cosmos, and we combine it with his baptismal theology, we can arrive at a fully ecological conception of baptism. If baptism is an eschatological sign of the coming Reign of God, then it is also the sign of a cosmic redemption, because the horizon of the coming Reign of God against the background of Christ's bodily resurrection is none other than the new creation. If baptism is a sign bound up with the whole eschatological history of Christ—a history of the baptized, crucified, resurrected, and coming Christ, and a history embedded in the Trinitarian history of God—then it is a sign for the baptized of a participation in that cosmic-divine history that sets the direction of a person's life in a messianic community

---

60. Ibid., 229.

and toward the future Reign of God.⁶¹ If baptism is a sign bound up with the whole eschatological history of the Holy Spirit—a story of the Sprit's leading and equipping Christ in His mission and resurrecting the dead Christ into an eschatological life and being sent through Christ to be poured out on every flesh, mediating the *Eschaton* and the present time with His/Her transforming power—then it must be a sign of cosmic transformation and rebirth in the life-giving Spirit after the rebirth of Christ. In this way, the eschatological sign of baptism in Moltmann's early ecclesiology can be re-identified as a sign of cosmic redemption in the context of eschatological Christology and eschatological pneumatology in the Trinitarian history of God.

THE EUCHARIST AS AN ANTICIPATED CELEBRATION OF THE NEW CREATION

The Eucharist has occupied the most important place in the Christian church, for better or worse. The appropriate theological understanding of the Eucharist based on eschatological understanding of the church itself and worship, therefore, is essential for today's Christians as well as the whole planet. Discussion on the Eucharist is, however, very difficult and sensitive. First, it was one of the most debated issues in the 16th century Reformation, finally dividing Lutheran tradition and the Reformed tradition. Then, subscribing to a relatively narrow range of positions on the doctrine of the Eucharist is a way Reformed churches displayed their awareness of belonging to a common tradition.⁶² For these reasons, retrieving what is one of the most significant elements in the Reformed tradition and then re-considering it in our present context needs a special effort and agility in both historical accuracy and the current context. Doing so does not require a blind repetition of tradition but rather a critical love of tradition that involves both respectful appreciation of its relevance and strength in its particular context and at the same time a cautious discerning of its limit to reform it for today. For that reason, I will survey our inherited Eucharistic theologies of Calvin and Zwingli, as the founding theologians of Reformed tradition, as well

---

61. Miroslav Volf also argues that faith gives the believer a double inclusion into the fellowship of a faith community and the fellowship of Trinity. *After Our Likeness*, 173.

62. van der Borght, "Reformed Ecclesiology," 187.

as Luther, who is still important not only to Lutherans, but also to all Protestant Christians including Reformed Christians, to understand key issues in Protestant theology. Following this, I will present Moltmann's Eucharistic theology, and finally, I will evaluate Moltmann's Eucharistic theology as a Reformed theology.

Inherited Eucharistic Theologies and Cosmogenetic Thinking

I would argue that today's efforts for liturgical renewal should go deeper than selective appropriation among our inherited sacramental theologies, which have been formulated in a different world-view, to reach a cosmological understanding of our current reality. The scientific community has presented a fundamentally different vision of reality from the ones we have known. Instead of a static and harmoniously ordered universe with a hierarchical order of spirit over matter, today's scientific understanding presents us with an ever-evolving, open-ended universe in which new realities ceaselessly come into being. Some call this kind of reality cosmogenetic, emphasizing the dynamic and ceaseless changing of the space-time continuum.[63] In this all-encompassing reality, literally everything is inseparably related in an unimaginably complex web. By a profound interdependence and continuous interaction, everything is becoming something else, and therefore constitutes a continuous becoming. This understanding of the cosmos as an interrelated and constant becoming has yet to penetrate most people's mind. Nevertheless, it is also true that this new scientific consensus is already a leading paradigm of contemporary cosmological thinking.

Theology cannot ignore this shift in cosmology and repeat traditional teachings formulated using the old paradigm. Instead, theology should strive to understand the implications of the new understanding of reality and express anew our understanding of God, the creation, and the relation of the two in light of contemporary thinking.

Liturgical renewal should be based on this new theological understanding, which again is based on a new theological engagement with the contemporary understanding of reality. Liturgy is expressive of our understanding of God and the relationship of God with us. Without changing our theological ideas, no amount of natural setting and emphasis on the value and beauty of nature in worship will match the se-

---

63. Berry and Swimme, *Universe Story*, 2–4, 71–78.

riousness of our time's colossal ecological demand. Although the use of as many natural elements as possible and exposure to nature as often as possible could be part of a more fundamental re-envisioning of liturgy, they cannot be considered an adequate strategy for liturgical renewal.[64]

This need to reconsider our inherited theologies in light of a new cosmogenetic paradigm leads us to re-evaluate the Eucharistic theologies of Luther, Zwingli, and Calvin in an ecological light and re-appropriate valuable theological insights for the construction of today's viable ecological theology. I choose these three Reformers because Zwingli and Calvin represent Reformed tradition and Luther, as a precedent Reformer, was the primary reference for these theologians. A thorough historical study of the Reformers' Eucharistic theologies, however, is beyond the scope of this thesis. Thus, from an ecological vantage point, this thesis will look mainly into one aspect of these Eucharistic theologies: Where is the resurrected Christ in relation to the Eucharist?

Luther's Eucharistic Theory: Consubstantiation

For Luther, the resurrected Christ is "in, with, and under" the bread and wine at every Eucharist. This position of Luther's in the context of sixteenth-century debate is called *consubstantiation*. Although he rejects the Catholic doctrine of transubstantiation, his idea regarding the mode of Christ's presence strongly emphasizes the real presence of Christ wherever the Eucharist is administered, instead of the so-called "spiritual" presence.

Santmire maintains that Luther's teaching about the ubiquity of the risen and ascended Christ should be understood in light of his theology of the *immanence of God* in the creation.[65] According to Santmire, because Luther understands space primarily as God's and then the world's, Luther could argue that the whole of the godhead is present in a grain of wheat. Similar to Moltmann's panentheistic description, Luther envisions the immediate presence of God within and yet beyond the creation. In this connection, Santmire claims:

---

64. These less than adequate suggestions are actually the basic orientation of some recent books aiming at liturgical renewal: McCarthy, *Celebrating the Earth*; Pearson, *Making Creation Visible*; Foley, Cauchin, Hughes, and Ostdiek, "Case Study in Liturgical Ecology"; Fragomeni, "Liturgy at the Heart of Creation."

65. Santmire, *Nature Reborn*, 83.

> [W]hen Luther states, concerning the ascension of the risen Lord to the right hand of the Father, that "the right hand of God is everywhere" (*dextera dei ubique est*), the reformer is merely giving his rich theology of the divine immanence what is for him an appropriate christological concreteness.[66]

In the two nature Christology, Christ's human body achieved God's divine ubiquity through *communicatio idiomatum*. Apparently, Luther was thinking of this doctrine when he overlapped God's pervasive and enveloping presence with that of Christ's. Therefore, Luther could argue that this is not just memory or spiritual substitution of Christ or the spiritual body of Christ; the real *flesh* and *blood* of Christ is "in, with, and under" the bread and wine.

Is Luther's doctrine of consubstantiation anti-ecological? Calvin criticizes this doctrine as obscuring the corporeality of the resurrected Christ. Since the resurrected body of Christ is connected to His human body, Calvin could not agree that Christ could be in all places at the same time, as he saw Luther contending. Thus, Calvin asks: "If Christ's body is so multiform and varied that it shows itself in one place but is invisible in another, where is the very nature of a body? Which exists in its own dimensions, and where is its unity?"[67]

Apparently, Calvin is rejecting Luther's idea that Christ's human body can receive the divine attribute of ubiquity by means of *communicatio idiomatum* because, for Calvin, the finite being (Christ's human body) cannot contain infinity (*finitum non capax infiniti*).[68] For Calvin, if a body is ubiquitous, it is no longer a human body. In this respect, we could agree with Calvin that Christ's human body which Luther claims to be "in, with, and under" the elements, is surely hard to recognize as human. So far as it obscures the bodily nature of Christ's resurrection and, thus, the hope of our bodily resurrection, it is not ecologically sound.

But, what if we see this theology of Luther in light of the presupposition of his own time, namely the Great Chain of Being and the Gothic spirit? Whereas Gothic-spirited worship posited that the soul

---

66. Ibid., 84.

67. *Institutes* IV.Xvii.19 cited in Rice and Huffstutler, *Reformed Worship*, 68.

68. This phrase is not Calvin's, but coined by later exponents of the Reformed doctrine. Richard A. Muller argues that this phrase is better rendered "The finite is unable to grasp the infinite." Muller, *Christ and the Decree*, 20.

ascended to the spiritual realm, leaving behind the material world, Luther's sacramental theology suggests the opposite direction in the vision of salvation. For Luther, salvation is possible not because we ascend above nature, but because God in Christ descends to this material world and fills it—first in the Incarnation and second in the mystery of the Eucharist. By eating and drinking the bread and wine, Christians are united both with the body and spirit of Christ. In this sense, Luther's Eucharistic theology countermands the anti-ecological Gothic spirit. In fact, the "condescension of God" for human salvation was already Luther's central theological theme.[69]

Therefore, in light of Luther's idea of the immanence of God and Christ, as well as his emphasis on human solidarity with nature,[70] his doctrine of consubstantiation could be understood as a mental picture of an ecological redemption, in which God descends to permeate and envelop the creation.

Zwingli's Eucharistic Theory: Transubstantiation of the People

For Zwingli, the resurrected Christ is in heaven, at the right hand of God the Father, not "in, with, and under" the bread and wine. In this respect, Zwingli shares with Calvin the same position regarding the ubiquity of Christ: Christ cannot be brought down to earth and be everywhere simultaneously.

Zwingli and Calvin, however, differed as to whether or not Christ is really present and whether or not the Eucharist is a means of grace. Zwingli denied the real presence of Christ and considered the Eucharist a memorial, not a means of grace.[71] However rich, moving, and symbolic it might be, the sacrament is a thanksgiving to God, a pledge of faith, a memorial of Christ's death, but not something that conveys grace. Grace is given by the Holy Spirit to the human spirit of the believer, but there is no connection with the bread and wine.[72] The language of memory replaces the language of real presence.

What really happens, then, at the Eucharist? If Christ is in heaven, and the bread and wine does not transform or transmit, what significance does this sacrament have? Zwingli states that its purpose is the

---

69. Santmire, *Nature Reborn*, 85; Idem, *Travail of Nature*, 127.
70. Ibid., 128–31.
71. Vischer, *Christian Worship in Reformed Churches*, 21.
72. Ibid., 14.

transformation of people, rather than the transformation of bread and wine. By consecrating the Communion in memory of Christ, it is not the elements but the people gathered that are transubstantiated into the body of Christ.[73]

Could Zwingli's doctrine of transubstantiation of the people be considered ecological? We must test Zwingli in the same beaker as Luther. As long as Zwingli's sacramental theology emphasizes and preserves the bodily nature of Christ's resurrection, it is potentially ecological because the resurrected body of Christ is an embodied promise of God's ecological redemption.[74] But, in light of Zwingli's understanding of the "spiritual" as non-material,[75] his sacramental theology might be understood as quite a-cosmic and anti-ecological. His fear of mingling spiritual truth with material elements signifies and inculcates, both for him and attending congregations, contempt of the material cosmos in which we move and breathe. Zwingli's sacramental theology is consonant with the Gothic spirit, in which salvation means rising above nature to ascend for a "spiritual" union with God. For Zwingli, then, the Eucharist is a dualistic vision of salvation. While we view this as being transformed collectively or individually into something worthy of salvation, we should not get distracted by the earthly dimensions of the bread and wine and what they are a part of; instead, we should look to heaven, where the crucified and resurrected Christ dwells. What, then, is the significance of Christ's bodily nature if He remains in heaven while we do everything to forget and escape our bodily form to reach the spiritual realm? Zwingli's idea of the transubstantiation of the people, however, carries great potential for our understanding of worship and its overflowing ecological mission.

Calvin's Eucharistic Theory: Spiritual Presence

For Calvin, the resurrected Christ is in heaven, at the right hand of God the Father. This means the resurrected *body* of Christ is not present at the Eucharist. With Zwingli, and against Luther, Calvin maintained that the body of the resurrected Christ is still a human body and is not made

---

73. Santmire, *Nature Reborn*, 88.

74. Lee, "Ecological Significance of the Resurrected Body of Christ," 13–16, 20–23.

75. Vischer, *Christian Worship in Reformed Churches*, 11. "One of Zwingli's favorite verses was John 6:63 (The Spirit gives life, the flesh profits nothing"), and he moved toward a very strong emphasis on the spiritual, where *spiritual* could sometimes mean the opposite of material." Moltmann also criticises Zwingli's concept of spirit as Platonic. *Church in the Power of the Spirit*, 252.

ubiquitous in every celebration. Calvin affirmed the real presence of Christ's body and blood, but he insisted it is spiritual, not physical or local.[76] However, Calvin differed from Zwingli in calling the real presence of Christ at the Eucharist a *spiritual presence.*

What is Calvin's motive in this argument? One may interpret Calvin's position as the result of a careful search for "a middle way between overidentification of signs and what it symbolised, on the one hand, and too great a divorce between them, on the other."[77] But in an ecological connection, Calvin's concern and reasoning was to oppose the reduction of Christ's resurrected body to a mere idea or a concept[78]; he wanted to preserve the bodily nature of Christ's resurrection, which is the foundation of our resurrection hope.[79]

If Christ cannot be brought down to us, how could Calvin still maintain that the Communion is a mystery and that Christ is present? Calvin claimed that it is the work of the Holy Spirit who raises us to Christ's presence when we eat the bread and drink from the cup. The Spirit unites Christ Himself with us and we participate mysteriously in the very being of Christ.[80]

Does Calvin's doctrine of the spiritual presence emphasize and preserve the bodily nature of Christ's resurrection to the point that it could be considered ecological? As long as this sacramental theology emphasizes the bodily nature of Christ's resurrection, it can serve as a strong reminder of a Christian ecological hope of a new heaven and new earth. On another level, however, Calvin's sacramental theology illustrates and enacts a vision of human salvation in terms of *leaving the earthly reality behind* and *ascending to heaven*, where the blissful union with Christ occurs. Therefore, we must ask: If Christ's human body stays in heaven and the Holy Spirit spiritually unites us with that body, what is the hope of salvation for unspiritual earthly realities? If our resurrection happens in heaven in the same way as the bodily resurrection of Christ, what physical / material environment will we have there for our body? What was the purpose, then, of God's gracious condescension? Is it not reasonable to say that it was to prepare us for our final escape from this

76. McKee, "Reformed Worship in the Sixteenth Century," 21–23.
77. Rice and Huffstutler, *Reformed Worship*, 67.
78. Ibid.
79. Ibid., 68.
80. Ibid.

world? Is not Calvin's doctrine of the spiritual presence of Christ, then, severely anti-ecological?

As is clear now, the emphasis and preservation of the bodily nature of Christ's resurrection is not enough upon which to base an ecological doctrine. Not surprisingly, Calvin's eschatology almost lacks the theme of "new heaven and new earth," focusing on the anthropocentric theme of the bodily resurrection of humankind.[81] This kind of ecological contradiction in Calvin's position—that is, the coexistence of the hope of bodily resurrection and lack of hope of ecological salvation—might come from his failure to perceive the fundamental connection and solidarity of humans with the earth,[82] whereas Luther apprehended and powerfully emphasized human solidarity with nature.[83]

Moltmann's Eucharistic Theology: Eschatological Presence

Moltmann expresses Christ's presence in the Eucharist in a very different way: the resurrected Christ who was crucified and is to come, is present in the feast of the Lord's Supper. To clarify further, the risen Christ is not localized "in, with, and under" the elements, nor is He considered to be distant or far away; the risen Christ is among the group of celebrants. This understanding of Christ's presence is different from any of the previously discussed sacramental theologies in that Moltmann proposes a shift from a spatial to a temporal and relational concept of presence.[84] If we force the question concerning the body of the resurrected Christ in this case, Moltmann would say that Christ is present in His whole being: crucified, resurrected, and now on His way to eschatological consummation.

---

81. Santmire, *Travail of Nature*, 126. "In his thoughts about the final fulfillment, although he holds to the traditional idea of the renewal of all things, he self-consciously avoids talking about the new heavens and the new earth. We can know very little about this, he says. Rather, he focuses all his discussion anthropocentrically on the theme of the bodily resurrection of the human creature, and related doctrinal issues."

82. This is probably due to his understanding of human beings as spirit and body, a typical understanding in the frame of the Great Chain of Being.

83. Santmire, *Travail of Nature*, 132.

84. *Church in the Power of the Spirit*, 253. Also, 155: "We must learn to think in a new way here: not—Christ is present in the feast here or there, but—the feast is held in his presence and carries those who partake of it into the eschatological history of Christ. . . ."

Moltmann criticizes Zwingli as only remembering the past and failing to perceive the presence of the crucified one in the Spirit of the resurrection; Moltmann also criticizes Luther as failing to properly relate the presence of Christ to His past event at Golgotha, blurring the significance of historical redemptive events in Christ's history.[85] Moltmann thinks the one-sidedness of these views is basically due to the use of spatial concepts, and proposes a temporal formula of Christ's presence: "In the one who is to come the one who died for us is present in our midst."[86]

By way of this temporal expression of Christ's presence, Moltmann wants to express the meaning of the Lord's Supper with historical concreteness and eschatological orientation:

> In the eschatological sense, the one who was once crucified on Golgotha is now himself present with the power of his suffering and the fruit of his death, in his giving of himself for many. This happening in the past is not a past event; it is an event which liberates, opens up the future, and therefore determines the present. In the temporal sense, the crucified Jesus is present as the One who is to come in the Spirit of the new creation and final redemption.[87]

In this way, Moltmann brings together Christ's past with its historical tangibility, as well as the eschatological direction aiming at the new creation, in order to formulate a temporal expression of the presence of Christ in the Eucharist.

Evaluation: Moltmann's Eucharistic Theology as an Eschatological Re-Appropriation of Reformed Eucharistic Theologies

So far, I have analyzed four theologians' views in terms of the presence of Christ at the Eucharist. Today, the theologies of Luther, Zwingli, and Calvin are important references for Reformed Christians when it comes to sacramental theology. We need to pay attention to those theologies in terms of ecological legacy: what ecological impact would they have had in their time and how valid are they currently?

---

85. Ibid., 253. "If we understand Christ's presence in the Lord's Supper along the same lines as the incarnation, then the christological difference between what happened on Golgotha and what happens on the altar can be easily overlooked."

86. Ibid., 254.

87. Ibid.

Luther's sacramental theology runs the risk of blurring the hope of a bodily resurrection, which could serve the ecological hope for the world. At the same time, however, his theology contains a merit that strongly emphasizes God's redemptive down-moving grace instead of a vision of deserting the earthly dimension to despair. Calvin and Zwingli, on the other hand, seem to be ecological because of their emphasis and preservation of the bodily nature of Christ's resurrection. Upon closer scrutiny, however, their theologies might represent a Gothic spirit, which encourages looking up to the spiritual realm, beyond the earthly and material reality. Here, however, it is important to note that the presupposed ontological / cosmological concept for the Reformers is the Great Chain of Being, in which spatial thinking determines spirituality. Therefore, the identification of the whereabouts of Christ's body was pursued in a vertical dimension, in terms of ascension of the soul above nature and the condescension of deity from heaven. Today's different understanding of reality requires a different understanding in theology. In today's cosmological scheme of *cosmogenesis* (an evolutionary universe), the unchanging, stable universe against which these Eucharistic ideas were developed does not exist. This cosmological presupposition requires a new kind of Eucharistic thinking, one that takes the constant becoming of space-time into account.[88]

Liturgy is closely related to theology and theology is based on a cosmological understanding. This is why liturgical renewal is not easy. Construction of ecological worship needs more than simple tactics, such as including more reference to nature, more use of green elements, and more appeals to stewardship, although all these things are helpful. Today's ecological liturgy should be based on today's best theological and cosmological understanding so that it can sink into the depths of the worshipper's psyche.

As noted earlier, the Great Chain of Being, dominated by spatial thinking, is the presupposed ontological / cosmological concept for the Reformers, and to a certain extent those of us in Western culture now. Today's scientific understanding is that the universe is an ever-

---

88. I do not mean that scientists have finally agreed on a certain cosmological theory, nor that theology always needs to be in perfect harmony with contemporary science. But the evolutionary universe is the basic consensus of many leading scientists today and is increasingly becoming our world view. Theology should maintain a dialogue with science, not simply ignoring this change in cosmology in articulating the relationship between God and the world, between humans and the rest of the creation.

evolving, open-ended universe, a *cosmogenesis* that emphasises the dynamic and ceaseless changing of the space-time continuum. In this regard, Moltmann's idea of the Eucharist is uniquely helpful, in that he introduces a new way of thinking about Christ's presence, not in terms of location in a static vertical universe but in terms of space and time by way of *anticipation*. By saying "the presence of the One who is to come," Moltmann's theology succeeds in introducing the presence of an eschatological new heaven and new earth without losing the historical concreteness of memory.

If an ecological liturgy is one that can evoke and reinforce the eschatological hope of a holistic future, Moltmann's idea of the presence of Christ at the Eucharist as an anticipation of that holistic future could be a strong ecological sacramental theology. First, when it comes to the bodily nature of the resurrected Christ, Moltmann does not hypothesise an unrecognizable spiritual entity as a substitute for the resurrected body of Christ; and thus the resurrected body of Christ can serve as the embodied ecological promise. Second, the presence of the One who is to come really signifies the anticipation of the eschatological kingdom of God, which, for Moltmann, is nothing other than the new heaven and the new earth. Thus, this presence of Christ does not divert our attention from historical and material realities, but leads us to a mission to transform *this* world. Third, his shift of theological attention, from the location of the body to the time dimension of the present Christ, helps us pay attention to the holistic, festive nature of the Eucharist; that is, the coming together of His people, biophysical elements, and divinity to celebrate eschatological consummation of the creation. This shift might enthuse people with the hope for an ecological redemption in which both history and nature reach their culmination with peace and justice.

Finally, Moltmann's eschatological understanding of Christ's presence in the Eucharist—that is, an eschatological understanding of the Eucharist, as an anticipated feast of the congregation in the presence of the crucified and resurrected Christ—could be viewed as a re-appropriation of Reformed sacramental theology. First, unlike Zwingli's one-sided spiritualistic view on the presence of Christ in the Eucharist, Moltmann's view attempts to preserve the corporeality of the resurrected One. Instead of too much attention on the elements as in Luther, Moltmann correctly emphasizes the anticipatorily festal nature in the presence of the Christ, who is the inaugurator and bringer of the new creation. In this regard, Moltmann's temporal expression could be one appropriation of Calvin's argument about Christ's presence in an eschatological light, as well as in today's cosmogenetic paradigm. Second, in the anticipated feast, Moltmann appropriates Zwingli's idea of transubstantiation of the people. In the presence of the crucified and resurrected One, the congregation glimpses into the future of the bodily transformation, the future of a healed and glorified cosmos, a world infused with peace and justice, rejoicing in the perichoretic fellowship in God.[89] This religious experience made possible by the eschatological dimension of the Holy Spirit might have been what Zwingli intended to express with the transubstantiation of the people, but was unable to articulate because he was trapped in dualistic spatial thinking. Third, Moltmann's concept of a perichoretic relationship between Christ and the Holy Spirit, or his employment of pneumatological Christology and christological pneumatology, helps us to better understand the Reformed claim of Christ's presence in the Eucharist. If Christ is perichoretically present in the work of the Holy Spirit that makes present the cross of Christ and the new creation inaugurated by Christ's resurrection, then the *living* Christ is present at the table, communicating Himself to the faith community through and with the elements. In that way, the Spirit-filled community at the table, that is, the faith-filled, hope-filled, and love-filled community is also *in Christ*. The spiritual presence of Christ is His perichoretic presence in the Holy Spirit, in His/Her Eucharistic work of presenting the eschatological feast of Christ. Therefore, it could be argued that Moltmann's Eucharistic theology is an eschatological re-appropriation of Reformed Eucharistic theology in the context of today's cosmogenetic understanding.

CONCLUSION

For Moltmann the church is an eschatological community created by the Holy Spirit to be a witness and tool in the process of the eschatological glorification of the Son and the Father. As such, the church mediates the present reality of which it is part and the eschatological reality that is coming. Furthermore, because the church is an anticipation of the *Eschaton*, worship, baptism and the Eucharist should also be seen in that light. The eschatological panentheistic reality expressed as the Reign of

---

89. For Eucharistic characteristic of worship, see above, "Worship and Mission."

God, the eschatological Sabbath, the new creation, and cosmic perichoretic indwelling of the Trinitarian God in creation, is mediated, becomes accessible and is celebrated in advance through worship, although it is always an anticipated feast, and thus, a partial participation.

Further, because worship contains the christological and Trinitarian remembrance and hope, it sets the everyday life of Christians in the context of the Trinitarian history, thus giving it a salvific meaning. Through worship, Christians grasp the grand perspective on the Trinitarian history spanning the remembrance (the life and passion of Christ) and the hope (the resurrection of Christ and the ensuing hope for the new creation). With this view of God's grand history, they can situate their current hope and struggle with the powers that be in the grand Trinitarian history; this enables Christians to make sense of their everyday life, grasping the meaning of daily struggle, pain and hope.

The anticipatory nature of Christian worship with remembrance and hope gives Christian worship a missional character. The hope and the anticipated participation in the eschatological panentheistic reality of God's Reign bring to light the inhumanity of human society and the exploitation and aggression on the natural world. In this way, Christians experience even more pain, restlessness, strife, and homelessness in this unredeemed state of the world order. From the tension experienced in worship between the anticipatory celebration of eschatological Sabbath on the one hand, and acute realization and feeling of the unredeemed state of the world on the other hand, springs the motivation and energy for the mission to protest and transform the world. Worship is and should be a smelting furnace in God's workshop, where worshippers are enlightened, inspired, immersed into an alternative vision and reality, and transformed into the image of God by the Holy Spirit. During and through this pneumatological experience, the *moral will* is formed and *moral imagination* is enabled.

This moral formation and ignition of moral imagination through worship does not always happen automatically. However, if we refuse to succumb to the pressure of today's society to compartmentalize our religion, to try to be palatable to religious consumers, and if we vow to base our worship once again on a sound theological understanding and eschatologically oriented liturgy, then, I would argue, worship can awaken the congregation to its eschatological significance and messianic mission. In so doing, it can become a true epicentre of transformation, a true smelting furnace to form a moral will, a true watchtower to envision an alternative way.

For Moltmann, baptism—as one of the two sacraments of the messianic community that anticipates the eschatological panentheistic reality in present history—can only be an eschatological sign, an eschatology put into practice that places the believer in the Trinitarian history of God by uniting the believer with Christ. As a synthesis of both historical and eschatological approaches to the significance of baptism, Moltmann points to its link both to Christ's death that took place once and for all, and to the resurrected Christ's presence in the Spirit which, as the first fruit of the new life, anticipates here and now the blissful common future that belongs to both the Trinitarian God and the creation.

Instead of a static and harmoniously ordered universe with a hierarchical order of spirit over matter, today's scientific understanding presents us with a cosmogenetic universe, an ever-evolving, open-ended space-time continuum in which new realities ceaselessly come into being. Instead of ignoring this shift in cosmology and repeating traditional teachings, theology should strive to critically understand the implications of the new understanding of reality and express anew our understanding of God, the creation, and the relation of the two in light of contemporary thinking. In this connection, Moltmann proposes to employ a temporal, instead of a spatial, formula of Christ's presence in the Eucharist: "In the one who is to come the one who died for us is present in our midst."[90] By using this temporal expression of Christ's presence, Moltmann wants to express the meaning of the Lord's Supper with historical concreteness and eschatological orientation. For Moltmann, the resurrected Christ who was crucified and is to come, is present in the feast of the Lord's Supper. To clarify further, the risen Christ is not localized "in, with, and under" the elements, nor is He considered to be distant or far away; the risen Christ is among the group of celebrants.

Moltmann's proposed shift in Eucharistic theology, and his perspective on the presence of Christ at the Eucharist as an anticipation of the eschatological future could be a strong ecological sacramental theology, especially for Reformed Christians. First, when it comes to the bodily nature of the resurrected Christ, Moltmann does not hypothesize an unrecognizable spiritual entity; and thus the resurrected body of Christ can serve as the embodied ecological promise. Second, the presence of the

---

90. Ibid., 254.

One who is to come signifies the anticipation of the eschatological kingdom of God, which, for Moltmann, is the new heaven and the new earth. Thus, this presence of Christ does not divert our attention from historical and material realities, but leads us to a mission to transform this world. Third, his shift of theological attention, from the location of the body to the time dimension of the present Christ, helps us pay attention to the holistic, festive nature of the Eucharist. This new understanding of the Eucharist could be viewed as an eschatological re-appropriation of the Reformed tradition. This shift might enthuse people with the hope for an ecological redemption in which both history and nature reach their culmination with peace and justice.

Now we have come to a point where we need to examine our liturgical practice in light of the eschatological panentheistic theology and envision an ecological liturgy that is grounded in the core understanding of Christian faith, is relevant for overcoming the current crisis and is striving toward an ecological era.

5

Envisioning an Ecological Reformed Worship

Theological Principles and Suggestions for an Ecological Reformed Worship Based on Moltmann's Eschatological Panentheism

> This life is not a state of being righteous,
> but rather, of growth in righteousness;
> not a state of being healthy, but a period of healing;
> not a state of being, but becoming;
> not a state of rest, but of exercise and activity.
> We are not yet what we shall be, but we grow towards it;
> the process is not yet finished, but is still going on;
> this life is not the end, it is the way to a better.
> All does not yet shine with glory;
> nevertheless, all is being purified. (2 Corinthians 3:18)
> ~ Martin Luther, Germany (1483–1546)[1]

SO FAR I HAVE shown that Moltmann's ecological theology is not only a viable Christian ecological theology but also has many features that grow out of the Reformed tradition, both in deference to, and critical reappropriation of it. With its overall eschatological directions expressed in Trinitarian history toward its glorification in the cosmos and inner relations among many theological themes such as Trinity, Christology, pneumatology, and ecclesiology, Reformed Christians could harness Moltmann's theology more vigorously to cope with today's ecological

---

1. Klug, *All Will Be Well*, 37.

crisis and, in the long run, to cultivate a moral will to fight against the current destructive path humanity is taking and a moral imagination to envision alternative ways to think and live.

In that regard, worship can play a key role. Although no one would deny the profound reciprocity and a mutual influence between theology and worship, a conscious effort should be made so that a theology so urgently needed for today may become more explicitly expressed liturgically.

This thesis does not strive to be a study in liturgy. My hope in this chapter is to open a dialogue with liturgical study by emphasising the importance of Moltmann's ecological theology for today and making some suggestions based on his theology. I want to complement this approach by adding some examples that are already in use in Reformed churches in Canada. Even though those examples may not always exactly express Moltmann's theological concepts, they might invite other liturgical scholars and creative artists to engage their creativity to envision more ecological Christian worship.

RE-ENVISIONING THE RELATIONSHIP BETWEEN GOD AND THE WORLD: TRINITARIAN-PNEUMATOLOGICAL IMAGE OF CREATION

In chapter 1, I laid out standards for a viable Christian ecological theology as re-envisioning the relationship between humans and the world and re-envisioning the relationship between God and the world. Most of chapter II has to do with the latter: Moltmann's concept of the Trinitarian history of God, the Trinitarian doctrine of creation, and the glorification of God in the cosmos as the climax of Trinitarian history. In this re-envisioning, the most crucial element in the picture has to do with the Trinitarian-pneumatological image of creation expressed in chapter 2.

This pictorial image, with all its limitations, conveys the panentheistic embrace of the cosmos by the Trinity as well as its eschatological reserve. As I argued in chapter 1, theology informs the most fundamental understanding of our reality, and a one-sided emphasis on God's transcendence over against creation continues to have a detrimental effect on our ecological way of thinking and living. This panentheistic understanding, then, should be more explicitly translated and expressed in liturgical forms. Translating this theological concept into a liturgical formula, as well as into lyrics of hymns, could foster a profound ecological spirituality. Emphasis on the Holy Spirit as the ontological bond in the context of the Trinitarian history could enhance the sense of the intrinsic value of creation as God's baby, cared for and nurtured toward maturation, without losing the balance between the immanence and transcendence of God.

The Spirit as the Ontological Bond in Trinitarian Creation and Sustenance

The dual reality of "already" and "not yet"—God's presence and God's absence in creation, and in the church—is mind-challenging and yet quite familiar to Christians. Worshippers need a suitable conceptual image that not only makes sense of this conundrum and deeply felt reality, but also expresses the worshiper's irrepressible yearning and prayer emanating from that reality.[2] Moltmann's use of *zimsum* to describe creation in the feminine imagery of conception and birthing, as well as the Spirit's umbilically nurturing connection, is surely something to liturgically harness for worshippers, who are confronted with the daily abuse and exploitation of the creation, and yet need hope in God's eschatological redemption.

The christological themes of resurrection and re-birth have a strong connotation for ecology. The Nicene description of the Holy Spirit as Life-giver is definitely in line with this image. Phrases such as "breath of life,"[3] "root of life,"[4] "nurture all who seek rebirth,"[5] "God of Hovering Wings," "Womb" and "birth of time"[6] might help envision such imagery. A song in *Voices United* has a special relevance in this regard:

> Mothering Spirit, nurturing one,
> in arms of patience hold me close,
> so that in faith I root and grow
> until I flower, until I know.[7]

2. Saliers, *Worship as Theology*, 21–24.

3. Edwards, *Breath of Life*; *Book of Praise*, 513; United Church of Canada, *Celebrate God's Presence*, 59, 276.

4. Presbyterian Church in Canada, *Book of Praise*, 513.

5. United Church of Canada, *More Voices*, no. 10.

6. Presbyterian Church in Canada, *Book of Praise*, 396.

7. United Church of Canada, *Voices United*, no. 320.

The Holy Spirit as the nurturing Mother must be, in the first place, a life-giving (or resurrecting) energy. In this connection, phrases that have pentecostal connotations such as the Wind,[8] and "source of vital energy"[9] might need to be used more frequently.

Another example of the expression for the umbilical nurturing of the cosmos by the Holy Spirit can be found in the following prayer:

> We bless you, for your Spirit wraps us in your presence,
> drawing us closer in community.[10]

The Glorification of God in the Cosmos as the Climax of Trinitarian History and Creation's Intrinsic Value in Light of Eschatological Panentheistic Redemption

> In the breaking of this bread,
> we remember God's creation
> and the groaning of this planet earth,
> bruised by thoughtlessness and greed.
> We pray for the day when the earth in all its fullness
> will reflect the glory of its Creator.[11]

Reformed ecological spirituality springs from God's covenantal love toward the creation. God's loving intention and purpose toward the creation, which will come to fruition at the eschatological redemption of the whole cosmos, can help us overcome a dualistic worldview and world-deserting eschatology and help us discover and appreciate the intrinsic value of fellow-creatures apart from their alleged utility for humans.

In this regard, eschatology should be revealed to today's worshippers. Although there are many songs and prayers that mention creation's beauty, there are not enough that put the creation in an eschatological light. Consequently, these songs and prayers often fail to address the current suffering, threats and dangers that haunt the creation and indulge in a one-sided praise of creation in the beginning. For example:

> Sing praise to God where fishes swim and birds fly in formation,
> where animals of every kind diversify creation.

---

8. United Church of Canada, *More Voices*, no. 5.
9. Presbyterian Church in Canada, *Book of Praise*, 513.
10. United Church of Canada, *Celebrate God's Presence*, 280.
11. Ibid., 300.

All life that finds its home on earth is meant to be respected;
let nothing threaten for base ends, what God through grace perfected.[12]

Although it expresses contemporary ecological concern in the latter part, it does not go so far as to emphasize its eschatological grace through which it will participate in the glory of God. The communion prayer I cited at the beginning of this section, however, attempts to address the eschatological hope of the earth as well as the current sufferings of the creation.

RE-ENVISIONING THE RELATIONSHIP BETWEEN HUMANS AND THE WORLD: WORSHIP AS AN ANTICIPATED FEAST OF THE NEW CREATION

Re-envisioning the relationship between humans and the world is another indispensable element of ecological worship for today. In this area, the church and worship become important. Moltmann's understanding of the church as a messianic community that anticipates the eschatological redemption here and now, and his understanding of worship as an anticipated feast of the eschatological new creation, should play a key role in moulding an ecological ecclesiology and an ecological missiology.

The New Creation as the Broadest Concept of Redemption

Instead of anti-ecological concepts of nature-ignoring, world-discarding eschatologies and theories of redemption, we should emphasize the new creation as the renewal of the old creation by the Holy Spirit in the Trinitarian history of God. Only within this concept of redemption, co-creaturehood of humanity with the rest of creation in the complicated web of life can make sense. Phrases such as "web of life"[13] could also be used more often to indicate the kinship between humanity and the whole creation.

Ecological responsibility of Christians in light of eschatological hope should be considered not as an extemporary concern of the church but an intrinsic Christian ethic. Songs and prayers that bring up the suf-

---

12. Presbyterian Church in Canada, *Book of Praise*, 416–17.
13. United Church of Canada, *More Voices*, no. 39.

fering of the creation need to be used more often in the eschatological hope we have in God.

Worship as an Anticipated Feast

The understanding of Christian worship itself cannot be over-emphasized for ecological spirituality. The eschatological significance of Christians gathering for worship as well as its sacraments of the Eucharist and baptism should be more fully expressed in liturgy.

The Eschatological Significance of Sunday Worship

According to Nicholas Paul Wolterstorff, when the sixteenth century Reformers brought about the most radical liturgical reform that the Christian church had ever known, they saw themselves not as beginning over but as returning to the liturgy of the early church, as the word "re-form" suggests. With the exception of Zwingli, this reform largely meant recovering the Lord's Day worship liturgy in the early church, comprising both of word and sacrament.[14]

This constitution of Sunday worship is not a blind following of tradition but a re-appropriation of a finely-tuned Reformed emphasis on God's initiative grace and our receiving of that grace on the one hand and our response to God's grace on the other. Wolterstorff emphatically remarks:

> [T]he liturgy as the Reformers understood and practiced it consists of God acting and us responding in faith through the work of the Spirit. The controlling idea in Reformed worship is that God acts in worship and that we are not to hold God's action at arm's length but to appropriate them into our innermost being. Worship is a meeting between God and his [sic] people, a meeting in which both parties act—God as the initiator and we as the responders.[15]

In this regard, understanding of and emphasis on the eschatological significance of Sunday worship should be the undercurrent of all liturgical awareness because, as this thesis strives to show, eschatology emphatically expresses God's initiative redemptive work and invites our

---

14. On detailed discussion about the early liturgy and later development as well as the Reformation of the liturgy in the sixteenth century, see Wolterstorff, "Genius of Reformed Liturgy." Online: http://www.reformedworship.org/magazine/article.cfm?article_id=24.

15. Ibid.

responsive participation in the future liberation and healing by means of anticipation. As Don E. Saliers argues, "At the heart of all Christian prayer and worship is the cry for God's will and covenant promises in Jesus Christ to be made real."[16]

The following citation from *The Book of Common Worship*, a liturgical guide for the Presbyterian Church in Canada, provides not only a good exposition of the eschatological significance of Sunday worship but also a good example of how we can continuously inculcate ministers and worship planners to do their work in accordance with the eschatological nature of worship.

> For in early Christians, the first day of the week was of special significance. On the first day of the week God began creation, calling light out of darkness. On the first day of the week the Lord Jesus Christ was raised from the grave, his ministry was validated, the powers of death were defeated, the promise of life eternal was assured. It was a day of new beginning, a day to celebrate the new covenant. The day of resurrection became known as "the Lord's Day." Each Sunday became a little Easter, giving new shape and meaning to the seven days of work and holy rest, and eventually giving shape to the church's annual calendar of Christian festivals.[17]

The Anticipation of New Creation in Baptism

For Moltmann, baptism is an eschatological sign of rebirth. I laid out in chapter IV an understanding of baptism as a sign bound up with the whole eschatological history of Christ, a history of a baptized, crucified, resurrected, and coming Christ, a history embedded in the Trinitarian history of God.[18] If that is so, baptism is a sign for the baptized of participation into that cosmic-divine history that sets the direction of the person's life toward the coming future Reign of God. This eschatological ecological significance of baptism cannot be emphasized too much.

The cosmic dimension of baptism in light of eschatological anticipation, beyond a community celebration, has been part of a key ecumenical understanding:

---

16. Saliers, *Worship as Theology*, 49.
17. *Book of Common Worship*, 15.
18. This understanding of baptism is also expressed in World Council of Churches, *Baptism, Eucharist and Ministry*, baptism para. 3.

> Baptism initiates the reality of the new life given in the midst of the present world. It gives participation in the community of the Holy Spirit. It is a sign of the Kingdom of God and of the life of the world to come. Through the gifts of faith, hope and love, baptism has a dynamic which embraces the whole of life, extends to all nations, and anticipates the day when every tongue will confess that Jesus Christ is Lord to the glory of God the Father.[19]

This theological understanding of baptism should be more explicitly translated into liturgical formula.

The Eucharist as an Eschatological Feast

Concerning the Eucharist, Moltmann wants to move beyond the spatial thinking that determined the old debate over the elements and thinks in temporal terms to emphasise its eschatologically festive nature. Accordingly in Moltmann's formula, Christ's presence is "in the one who is to come the one who died for us is present in our midst."[20] In other words, the resurrected Christ is present among the celebrants of the Eucharist as the one who is to come in His eschatological glory, beaming out the light of the new creation. In this way, Moltmann expresses the meaning of the Eucharist with historical concreteness and eschatological orientation.

This theological understanding, suitable to today's cosmogenetic understanding of time and space, needs more explicit expression in worship. Worship needs to make explicit the Eucharist as an anticipated feast of eschatological renewal of the whole creation; worship needs to visualize the ecologically restored and consummated future of the cosmos. Only the foretaste of the coming Reign in the form of an anticipated feast and the embodied and already-partially-realized promise in the form of resurrected body of Christ can overcome the hope-crushing reality of our time.

The understanding of the Eucharist as a foretaste of eschatological renewal of creation is not new to ecumenical understanding:

> The eucharist opens up the vision of the divine rule which has been promised as the final renewal of creation, and is a foretaste of it. Signs of this renewal are present in the world wherever the grace of God is manifest and human beings work for justice, love

---

19. "Baptism, Eucharist and Ministry," baptism para. 7.
20. *Church in the Power of the Spirit*, 254.

and peace. The Eucharist is the feast at which the church gives thanks to God for these signs and joyfully celebrates and anticipates the coming of the Kingdom in Christ (I Cor. 11:26; Matt. 26:29).[21]

In line with this understanding of the Eucharist as an eschatological sign of a cosmic feast, phrases such as "foretaste of feast" and "God's renewal of all things" could be used more frequently.

Fostering Ecological Spirituality in Worship and Sending out for Ecological Mission

The eschatological understanding of Sunday worship and the sacraments in terms of cosmic renewal opens up a whole new range of possibilities for us to express ecological hope and zeal for ecological mission. Prayers should be raised for ecological sins and the undue sufferings of creation, and ecological hopes in God's grace should be expressed more often. Furthermore, this condensed experience in the worship of the suffering present and glorified future of creation should spill over to the rest of the week.

THE CHRISTOLOGICAL GROUNDING OF ECOLOGICAL SPIRITUALITY: THE CROSS AND RESURRECTION AS THE FOUNTAIN OF CHRISTIAN ECOLOGICAL HOPE

For Reformed Christians, Christ-centeredness is a non-negotiable theological criterion. Christ has always been the center of theological thinking and piety. In the area of ecological theology, Christ is still our righteousness, holiness and redemption, and thus, ecological hope should be found in Christ, who was crucified and risen and is to come. Moltmann's theological strength in this regard becomes manifest when he positively emphasises the cosmic dimension of the cross and resurrection of Christ with ecological sensitivity.

The eschatological-Trinitarian understanding of the cross opens up a spatial image of the Trinitarian embrace of the cosmos, thus making the cross the ontological foundation for ecological redemption. The bodily resurrection of Christ understood in a Trinitarian way represents an embodied promise of transforming the old creation into the new

---

21. "Baptism, Eucharist, and Ministry" Eucharist para. 22.

creation. In this christological scheme, the whole cosmos is taken into the Christ event and emerges with a pneumatological hope of the eschatological redemption and the ensuing task of the church for the here and now. The christological grounding of an ecological hope not only appeals to Reformed piety but also relates the mission of the church to a fundamental Christian hope and faith.

The Cross Event as the Embrace of the Cosmos in the Trinitarian Suffering

Hymns and prayers in the Reformed tradition that involve the cross event are invariably penitential; they petition for the cleansing of the soul, and rarely feature ecological underpinnings. These individualistic and penitential tendencies in the cross reflection, however, should shift to embrace the Trinitarian-cosmic dimensions of the cross event so that it may become the centre of Christian hope in this time of ecological crisis.

The Resurrection of Christ as the Inauguration of Cosmic Transition to the New Creation

The resurrection, for Moltmann, is the epistemological foundation for the cosmic hope since it is the first rebirth of Jesus by the Holy Spirit into eternal life of the new creation. From the resurrected body of Christ comes the Holy Spirit that renews the face of the earth. The resurrected Christ commissions the church to be the witness to, and anticipation of, the coming eschatological reality. In this context, liturgical texts should emphasize more vividly the eschatological cosmic significance of Christ's resurrection, not just the consolation of individuals and the church.

The bodily nature of Christ's resurrection is also crucial in this ecological understanding of Jesus' resurrection, constituting the first transition—not the discarding—of the old into the new transformed reality in which God will be glorified. The emphasis on God's covenant faithfulness as Creator, ensuring the continuity between the old and new creation, is not only in line with Reformed piety but also something that needs more urgent emphasis in face of today's ecological crisis. In the parable of the dying grain of wheat, the Gospel of John provides a good illustration of this transitional aspect of the resurrection, which would enhance our ecological sensibility instead of fostering a sense of total disruption and discontinuity in imaging the resurrection. After all, it is the crucified One that is resurrected and is to come. In this regard, phrases such as "the first born of the new creation,"[22] "the risen Christ who makes all things new,"[23] and additional imagery of the spring time renewal of life[24] can reveal the transitional aspect of Christ's resurrection and thus the ecological significance of it.

The Spirit of Resurrection as the Nurturing and Transformative Power Toward the New Creation

The Holy Spirit as Creator Mother is a mirror image of the present Christian experience of rebirth and healing amidst the brokenness of the world, just as the cosmic Christ is the mirror image of the resurrected Christ in an eschatological light. If the Holy Spirit coming from the risen Christ brings about healing and rebirth in the suffering and decaying present world, working as a nurturing and transforming power toward the new creation, the same Holy Spirit must be the creative Breath and Wind that brought about the initial creation. This pneumatological understanding equips and empowers the church for the ecological mission, both toward the preservation of what has been created and toward witnessing to what is to come in the transformation and consummation of the initial creation.[25]

Too many songs and prayers of the Pentecostal event are limited to its meaning for the church, far short of arousing ecological hope and energy for the ecological mission. The following hymn, however, partially demonstrates the transformative power of the Holy Spirit to renew the creation.

> 1. O Breath of life, come sweeping through us;
> revive your church with life and power.
> O Breath of life, come cleanse, renew us,
> and fit your church to meet this hour.
>
> 2. O Wind of God, come bend us, break us,
> till humbly we confess our need;

---

22. United Church of Canada, *Celebrate God's Presence*, 277.

23. Ibid., 279.

24. Ibid., 196. Traditionally Christian worship has steered away from these images due to fertility cults. But perhaps now is the time for Christian worship to reclaim the wonderfully beautiful demonstration of the Spirit's work in nature.

25. On the four activities of the Holy Spirit, see the section titled "The Eschatological Spirit and the Cosmic Spirit" in chapter 2.

> then in your tenderness remake us,
> revive, restore, for this we plead.
>
> 3. O Breath of love, come breathe within us,
> renewing thought and will and heart;
> come, Love of Christ, afresh to win us,
> revive your church in every part.[26]

Another critical point in nurturing a Christian ecological spirituality lies in not dimissing the Holy Spirit as just a cosmic spirit or an abstract principle and power inherent in the natural process. To do so blurs the Christian hope based on Jesus Christ. In this regard, it is crucial to understand theologically and express liturgically the close relation between the christological basis and pneumatological present. Therefore, it is necessary to relate the power and energy of the Holy Spirit toward transformation to the christological foundation as often as possible so that the Christian hope may not drift away from historical concreteness and remain tangible in christological forms.

Lastly, nurturing ecological spirituality includes bemoaning the dying and suffering creation due to our greed and negligence. The groaning of the Spirit with the creation in suffering (Romans 8), as well as the prayerful longing of the church for the renewal of the earth should be mentioned more often. Therefore, grief and sighing over the suffering and dying creation as well as an ecological petition for consummation of God's new creation should also be present in the worship.

CONCLUSION

In this chapter, I have expressed the need to translate Moltmann's ecological theology into liturgical forms to enhance the theological understanding of worshippers as well as to help form a moral will and ignite a moral imagination.

When it comes to re-envisioning the relationship between God and the world, the pictorial image of the Holy Spirit that umbilically connects and nurtures the creation to lead and invite it to the maturation and consummation is of crucial significance. Emphasis on the Holy Spirit as the ontological bond in the context of the Trinitarian history could enhance the sense of the intrinsic value of creation as God's baby, cared for and

---

26. Presbyterian Church in Canada, *Book of Praise*, 502.

nurtured toward maturation, without losing the balance between the immanence and transcendence of God. The christological-pneumatological themes of resurrection and re-birth also have a strong connotation for ecology. The Nicene description of the Holy Spirit as Life-giver is definitely in line with this image. Accordingly, phrases such as "breath of life," "root of life," "nurture all who seek rebirth," "God of Hovering Wings," "Womb" and "Birth of time" might help envision such imagery.

Reformed ecological spirituality springs from God's covenantal love toward the creation and thankfully we have a fair amount of prayers and hymns on the theme of creation's beauty. There is, however, not enough liturgical emphasis on the eschatological perspective on the creation. There is a lamentable shortage of prayers and hymns on the current sufferings and yearnings of the creation, as well as the eschatological redemption, through which the whole creation will participate in the future glory of God.

When it comes to re-envisioning the relationship between humans and the world, Moltmann's eschatological understanding of the church and worship should play a key role in molding a more ecological Christian liturgy. Instead of anti-ecological, nature-ignoring, and world-discarding eschatologies and theories of redemption, we should emphasize the new creation as the renewal of the old creation by the Holy Spirit in the Trinitarian history of God. On the basis of this, the eschatological understanding of the church and worship, including baptism and the Eucharist, as an anticipated feast and foretaste of eschatological renewal of creation, is not only in accordance to ecumenical theology but also a crucial nexus between ecology and Christian worship.

Based on such understanding and practice, Christian worship can and should function as the basis for ecological mission. The eschatological understanding of Sunday worship and the sacraments in terms of cosmic renewal opens up a whole new range of possibilities for us to express ecological hope and zeal for ecological mission. Prayers should be raised for ecological sins and the undue sufferings of creation and ecological hopes in God's grace should be expressed more often. This condensed experience in the worship of the suffering present and glorified future of creation should spill over to the rest of the week.

For Reformed sensitivity in relation to ecological spirituality, the Christian ecological hope should be firmly grounded on our faith in Jesus Christ. The eschatological-Trinitarian understanding of the cross opens up a spatial image of the Trinitarian embrace of the cosmos, thus making the cross the ontological foundation for ecological redemption. The bodily resurrection of Christ understood in a Trinitarian way represents an embodied promise of transforming the old creation into the new creation. In this christological scheme, the whole cosmos is taken into the Christ event and emerges with a pneumatological hope of the eschatological redemption and the ensuing task of the church for the here and now. The christological grounding of an ecological hope not only appeals to Reformed piety but also relates the mission of the church to a fundamental Christian hope and faith.

For that purpose, traditionally individualistic and penitential tendencies in the cross reflection of the cross should shift to embrace the Trinitarian-cosmic dimensions of the cross event so that the cross of Christ may become the centre of Christian hope in this time of ecological crisis. Liturgical texts should also emphasize more vividly the eschatological cosmic significance of Christ's resurrection, not just the consolation of individuals and the church. The bodily nature of Christ's resurrection is also crucial to instill in us the Creator's faithfulness because Christ's resurrected body constitutes the first transition—not the discarding—of the old into the new transformed reality in which God will be glorified.

The continuing nurturing and transforming work of the Spirit in the church as well as in the creation should also be liturgically emphasized. If the Holy Spirit coming from the risen Christ—the experience of the church—brings about healing and rebirth in the suffering and decaying present world, working as a nurturing and transforming power toward the new creation, the same Holy Spirit must be the creative Breath and Wind that brought about the initial creation. In this regard, it is important to relate the power and energy of the Holy Spirit toward transformation to the christological foundation. This pneumatological understanding equips and empowers the church for the ecological mission, both toward the preservation of what has been created and toward witnessing to what is to come in the transformation and consummation of the initial creation.

Conclusion

Christian Worship as Celebrating God's Eschatological Cosmic Perichoresis

REVIEW OF THE THESIS

THIS THESIS AIMS TO appropriate Jürgen Moltmann's ecological theology, in the framework of Christ-centered Trinitarian history, into worship, especially in a Reformed worship, in order to enhance ecological awareness and hope in and beyond the church. For that purpose, I structured the thesis in a way that Reformed principles and characteristics (chapters 1 and 5) enfold Moltmann's understanding of the Trinity, Christ and the church (chapters 2, 3, and 4 respectively), among which the christological grounding of ecological redemption stands at the center.

In chapter 1, I presented standards for a viable Reformed theology from characteristics and principles of Reformed worship and theological approach. I also defined the critical tasks of Christian ecological theology in terms of re-envisioning the relationship between humans and the world and between God and the world. Next, I showed that Moltmann's eschatological panentheism satisfies both sets of standards with a brief presentation of the characteristics, the orientations and inner-relations of his theology. Then, I emphasised the special relevance of an eschatological panentheism for today's Christian ecological theology from the Reformed perspective.

In chapter 2, I first presented the Trinitarian history of God as the overall framework of Moltmann's ecological theology. I explicated this concept in the context of the ecological cosmology to show the affinity between Moltmann's ecological theology and the newly emerging cosmogenetic cosmology. Using this concept of Trinitarian history as a basis, I presented Moltmann's doctrine of creation with a description of his panentheistic spatial thinking. In doing this, I made a special effort to reveal the role of the Holy Spirit in His/Her eschatologically bonding, nurturing and guiding connection. In the last part of this chapter, I presented the consummation of Trinitarian history as the glorification of God in the cosmos—God finds eschatological rest in the perichoretic indwelling in the creation, which is the redemption of all creation.

In chapter 3, I presented Moltmann's eschatological Christology in light of the Trinitarian history of God. The inseparable inter-relationship between Moltmann's Christology and his doctrine of the Trinity demands a close re-examination of the significance of the Christ event in light of the eschatological-Trinitarian outlook of the creation. In this re-examination, I presented Moltmann's understanding that the resurrection of Christ is the epistemological foundation for cosmic hope while the cross is the ontological foundation for ecological redemption.

In chapter 4, I articulated the meaning and purpose of the church and its worship in light of the eschatological-Trinitarian interpretation of the Christ event. In this perspective, I argued that the church emerges as an eschatological community grounded in the Trinity and its eschatological history; it is called to be a witness and tool in the process of eschatological glorification of the Son and the Father by the Holy Spirit. Furthermore, the anticipatory nature of Christian worship with remembrance and hope gives Christian worship a missional character. In other words, from the tension experienced in worship between the anticipatory celebration of eschatological Sabbath on the one hand, and acute realisation and feeling of the unredeemed state of the world on the other hand, springs the motivation and energy for the mission to protest and transform the world. In this context, I also examined and evaluated Moltmann's understanding of baptism and the Eucharist to appropriate it in a Reformed ecological worship.

In chapter 5, I expressed the need to translate Moltmann's ecological theology into liturgical forms to enhance the theological understanding of worshippers as well as to help form moral will and ignite moral imagination. I presented liturgical examples from Reformed churches in line with the eschatological-Trinitarian panentheism of Moltmann that could help us 1) re-envision the relationship between God and the world, 2) re-envision the relationship between humans and the world, and 3) emphatically appropriate the Christ event as the ground of Christian ecological spirituality.

TOWARD AN ECOLOGICAL CHRISTIAN WORSHIP CELEBRATING GOD'S ESCHATOLOGICAL COSMIC PERICHORESIS

In this final section, in spite of the risk of repetition with conclusions in each chapter, I would like to summarily highlight some of the most important themes and emphases that emerged in the thesis.

The "Already" and "Not Yet" of the Creation

Throughout this thesis, I emphasised the double status—"already" and "not yet"—of the creation in relation to God. This tension relates theologically to the concept of an eschatologically-oriented continuing creation and the umbilical connection of the Holy Spirit to the created world. In ecclesiological and missiological thinking, this tension constitutes the basis of a congregation coming together for worship, bringing both the suffering of the creation with expectation of renewed taste of new creation, and then sending out of the congregation into the world again for an ecological mission. The significance of this understanding lies in helping us avoid a one-sided protological emphasis on the beauty and perfection of the creation, ignoring the suffering both in human history and nature, as well as tapping into the ecological mission of the church.

Christ our Righteousness, Holiness, and Redemption

One of the core tenets of the thesis is that our ecological hope and missional energy for reconciliation and harmony with nature should come from the head of the body—Jesus Christ, who, as our righteousness, holiness and redemption, is the foundation for the new creation and thus for an ecological redemption. In this regard, ecological understanding of the Christ event in the framework of the Trinitarian history of God is crucial. When seen in this perspective, the cross is the division and unity event in the Trinity. Through the unity that the Holy Spirit brings to the division between the Father and the Son experienced on the cross, the whole Trinity embraces the totality of the created reality in the inconceivably deep and wide space created *in* the Trinity. Because Christ participated in the death of all finite living things, which again springs from and forms part of the physical and natural world, the whole cosmos is taken into the Christ event and emerges with a pneumatological hope of the eschatological redemption and the ensuing task of the church for the here and now. In this context, the resurrected body of Christ takes on a special ecological significance because it is the inaugural focal point, an embodied divine promise of eschatological transition and transformation. It is also an expression of the faithfulness of the Creator, who creates the new creation out of the old in a manner that suggests a transition rather than a radical discontinuity.

The Holy Spirit as the Umbilical Cord

When it comes to re-envisioning the relationship between God and the world, the pictorial image of the Holy Spirit, who umbilically connects and nurtures the creation to lead and invite it to the maturation and consummation, is of crucial significance. Emphasis on the Holy Spirit as the ontological bond in the context of the Trinitarian history could enhance the sense of intrinsic value of creation as God's baby, cared for and nurtured toward maturation, without losing the balance between the immanence and transcendence of God. In addition, the continuing nurturing and transforming work of the Holy Spirit in the church as well as in the creation should also be liturgically emphasised.

If the Holy Spirit coming through the risen Christ—the experience of the church—brings about healing and rebirth in the suffering and decaying present world, working as a nurturing and transforming power toward the new creation, the same Holy Spirit must be the creative Breath and Wind that brought about the initial creation. In this regard, it is important to relate the power and energy of the Holy Spirit toward transformation to the christological foundation. This pneumatological understanding equips and empowers the church for the ecological mission, both toward the preservation of what has been created and toward witnessing to what is to come in the transformation and consummation of the initial creation.

The Eschatological Understanding of the Church and its Worship Grounded in the Trinity and its History

The church of Jesus Christ is a creation by the Holy Spirit who moves toward the eschatological glorification of the Son and the Father. Thus, the messianic community bearing the name of Christ is from beginning to end an eschatological community. It anticipates, embodies, participates in, and witnesses to God's eschatological Reign. For that reason, the church mediates the present reality of which it is part and the eschatological reality that is coming. Furthermore, because the church is an anticipation of the *Eschaton*, worship, baptism and the Eucharist should also be seen in that light. Therefore, the eschatological panentheistic reality expressed as the Reign of God, the eschatological Sabbath, the new creation, and cosmic perichoretic indwelling of the Trinitarian God in the creation, is mediated, becomes accessible, and is celebrated in advance through worship, although it is always an anticipated feast, and thus, a partial participation.

Cosmic Horizon for Worship and Mission of the Church

I have pointed to the cosmic horizon in which Moltmann's Christology has been developed: the cross as the ontological foundation for ecological redemption and the resurrection as the epistemological foundation for cosmic hope. Correspondingly, for Moltmann, the eschatological task of the Holy Spirit aims to glorify the Son and the Father in the cosmos, through the process toward the *Eschaton* where the cosmos becomes the dwelling place for the Trinitarian God, participating in the fellowship of the Trinity. If we gladly acknowledge this cosmic horizon for Christology and pneumatology, then we must see Christian worship involving no less than the foreseeing and foretasting of the new creation in Christ through the Holy Spirit and thus participating in the suffering and joy of the Trinitarian mission process. These christological, pneumatological, and ecclesial understandings invite a doxology that far exceeds the boundaries of the church's and humanity's concern and expresses the cosmic dimension of Trinitarian redemption.

WORSHIP AS THE EPICENTER OF THE ECOLOGICAL MISSION

Because worship contains the Trinitarian remembrance and hope, it sets the everyday life of Christians in the context of the Trinitarian history. Through worship, Christians grasp the grand perspective on the Trinitarian history spanning the remembrance (the life and passion of Christ) and the hope (the resurrection of Christ and the ensuing hope for creation). With this view of God's grand history, they can situate their current hope and struggle with the powers that be in the grand Trinitarian history; this enables Christians to make sense of their everyday life, grasping the meaning of daily struggle, pain and hope. Furthermore, the tension in worship between God's anticipated perichoretic indwelling and glorification in the cosmos on the one hand and the present unredeemed reality on the other, can be translated into the church's missional energy for a mission that has a cosmic horizon. In this connection, worship can be seen as a smelting furnace in God's workshop where worshippers are enlightened, inspired, immersed into an alternative vision and reality, and transformed into the image of the missional perichoretic God who pursues the cosmic perichoretic indwelling.

The Importance of Ecological Liturgy for the Well-Being of the Creation

Based on such understandings and practice, Christian worship can and should function as the basis for an ecological mission. The eschatological understanding of Sunday worship and the sacraments in terms of cosmic renewal opens up a whole new range of possibilities for us to express ecological hope and zeal for an ecological mission. Prayers should be raised for ecological sins, and the undue sufferings of creation, and ecological hopes in God's grace should be expressed more often. This condensed experience in the worship of the suffering present and glorified future of creation should spill over to the rest of the week.

Bibliography

PRIMARY SOURCES

Moltmann, Jürgen. "The Alienation and Liberation of Nature." In *On Nature*, edited by Leroy S. Rouner, 133–44. Notre Dame: University of Notre Dame Press, 1984.

———. *The Church in the Power of the Spirit: A Contribution to Messianic Ecclesiology*. Translated by Margaret Kohl. New York: Harper & Row, 1977.

———. "Creation and Redemption." In *Creation, Christ, and Culture: Studies in Honor of T. F. Torrance*, edited by Richard McKinney, 119–34. Edinburgh: T. & T. Clark, 1976.

———. "'The Crucified God': A Trinitarian Theology of the Cross." *Interpretation* 26:3 (1972) 278–99.

———. *The Crucified God: The Cross as the Foundation and Criticism of Christian Theology*. Translated by R. A. Wilson and John Bowden. New York: Harper & Row, 1973.

———. "Christian Hope: Messianic or Transcendent: A Theological Discussion with Joachim of Fiore and Thomas Aquinas." *Horizons* 12 (1985) 328–48.

———. *God in Creation: A New Theology of Creation and the Spirit of God*. Minneapolis: Fortress, 1985.

———. "Christ in Cosmic Context." In *Christ and Context: The Confrontation between Gospel and Culture*, edited by Hilary Regan, Alan J. Torrance, and Antony Wood, 180–91. Edinburgh: T. & T. Clark, 1993.

———. "Christianity in the Third Millennium." *Theology Today* 51 (1994) 75–89.

———. *The Coming of God: Christian Eschatology*. Translated by Margaret Kohl. London: SCM, 1996.

———. "The Cosmic Community: A New Ecological Concept of Reality in Science and Religion." *Ching Feng* 29:2–3 (1986) 93–105.

———. "The Ecological Crisis: Peace with Nature." *Scottish Journal of Religious Studies* 9 (1988) 5–18.

———. *Experiences in Theology: Ways and Forms of Christian Theology*. Translated by Margaret Kohl. Minneapolis, Fortress, 2000.

———. *Experiences of God*. Translated by Margaret Kohl. London: SCM, 1980.

———. "The Fellowship of the Holy Spirit: Trinitarian Pneumatology." *Scottish Journal of Theology* 37 (1984) 287–300.

———. "The Future as a New Paradigm of Transcendence." *Concurrence* 1 (1969) 334–45.

———. *The Future of Creation*. Translated by Margaret Kohl. Philadelphia: Fortress, 1979.

———. *God for a Secular Society: The Public Relevance of Theology*. Translated by Margaret Kohl. Minneapolis: Fortress, 1999.

———. *God in Creation: A New Theology of Creation and the Spirit of God*. Translated by Margaret Kohl. Minneapolis: Fortress, 1993.

———. "God's Kenosis in the Creation and Consummation of the World." In *The Work of Love: Creation as Kenosis*, edited by John C. Polkinghorne, 137–51. London: SPCK, 2001.

———. *History and the Triune God: Contributions to Trinitarian Theology*. Translated by John Bowden. New York: Crossroad, 1992.

———. "Hope and History." *Theology Today* 25 (1968) 369–86.

———. "Human Rights, the Rights of Humanity and the Rights of Nature." In *Ethics of World Religions and Human Rights*, edited by Jürgen Moltmann and Hans Küng, 120–35. Philadelphia: Trinity, 1990.

———. "I Believe in God the Father: Patriarchal or Non-patriarchal Reference?" *Drew Gateway* 59:2 (1990) 3–25.

———. "In Search for an Equilibrium of 'Equilibrium' and 'Progress.'" *Ching Feng* 30:1–2 (1987) 5–24.

———. "Is 'Pluralistic Theology' Useful for the Dialogue of World Religions?" In *Christian Uniqueness Reconsidered: The Myth of a Pluralistic Theology of Religions*, edited by Gavin D'Costa, 149–56. Maryknoll, NY: Orbis, 1990.

———. "Is the World Coming to an End or Has Its Future already Begun? Christian Eschatology, Modern Utopianism and Exterminism." In *The Future as a God's Gift: Explorations in Christian Eschatology*, edited by David Ferguson and Marcel Sarot, 129–38. Edinburgh: T. & T. Clark, 2000.

———. "Liberating and Anticipating the Future." In *Liberating Eschatology: Essays in Honor of Letty M. Russel*, edited by Margaret A. Farley and Serene Jones, 189–208. Louisville: Westminster John Knox, 1999.

———. "The Liberation of the Future and its Anticipations in History." In *God will be All in All: The Eschatology of Jürgen Moltmann*, edited by Richard Bauckham, 265–89. Edinburgh: T. & T. Clark, 1999.

———. "The Life Signs of the Spirit in the Fellowship Community of Christ." In *Hope for the Church: Moltmann in Dialogue with Practical Theology*, edited by Theodore Runyon, 37–56. Nashville: Abingdon, 1979.

———. *Man: Christian Anthropology in the Conflicts of the Present*. Translated by J. Sturdy. London: SPCK, 1974.

———. "The Motherly Father: Is Trinitarian Patripassianism Replacing Theological Patriarchalism?" *Concilium* 143 (1981) 51–56.

———. "Perichoresis: an Old Magic Word for a New Trinitarian Theology." In *Trinity, Community, and Power: Mapping Trajectories in Wesleyan Theology*, edited by M. Douglas Meeks. Nashville: Kingswood, 2000.

———. "Reconciliation with Nature." *Pacifica* 5 (1992) 301–7.

―――. "Reflections on Chaos and God's Interaction with the World from a Trinitarian Perspective." In *Chaos and Complexity: Scientific Perspectives on Divine Action*, edited by John Robert Russell, Nancey C. Murphy, and Arthur R. Peacocke, 205-10. Vatican City: Vatican Observatory, 1995.

―――. *Religion, Revolution, and the Future*. Translated by Douglas Meeks. New York: Scribners, 1969.

―――. "Resurrection as Hope." *Harvard Theological Review* 61 (1968) 129-47.

―――. *Science and Wisdom*. Translated by Margaret Kohl. Minneapolis: Fortress, 2003.

―――. "Shekinah: the Home of the Homeless God." In *The Longing for Home*, edited by LeRoy S. Rouner, 170-84. Notre Dame: University of Notre Dame Press, 1996.

―――. "Some Reflections on the Social Doctrine of the Trinity." In *Christian Understanding of God Today*, edited by James M. Byrne, 104-11. Dublin: Columbia, 1993.

―――. *The Spirit of Life: A Universal Affirmation*. Translated by Margaret Kohl. London: SCM, 1992.

―――. "Theological Proposals towards the Resolution of the Filioque Controversy." In *Spirit of God, Spirit of Christ: Ecumenical Reflections on the Filioque Controversy*, edited by Lukas Vischer, 164-73. Geneva: World Council of Churches, 1981.

―――. *Theology of Hope: On the Ground and Implications of a Christian Eschatology*. Translated by James W. Leitch. Minneapolis: Fortress, 1967.

―――. "Theology of Mystical Experience." *Scottish Journal of Theology* 32:6 (1979) 501-20.

―――. "The Trinitarian History of God." *Theology* 78 (1975) 632-46.

―――. "The Trinitarian Personhood of the Holy Spirit." In *Advents of the Spirit: An Introduction to the Current Study of Pneumatology*, edited by Bradford E. Hinze &andD. Lyle Dabney. Milwaukee: Marquette University Press, 2001.

―――. *The Trinity and the Kingdom: The Doctrine of God*. Translated by Margaret Kohl London: SCM, 1981.

―――. "The Unity of the Triune God: Comprehensibility of the Trinity and its Foundation in the History of Salvation." *St. Vladimir's Theological Quarterly,* 28 (1984) 157-71.

―――. *The Way of Jesus Christ: Christology in Messianic Dimensions*. Translated by Margaret Kohl. Minneapolis: Fortress, 1993.

―――. "The World in God or God in the World?" In *God will be All in All: The Eschatology of Jürgen Moltmann*, edited by Richard Bauckham, 35-42. Edinburgh: T. & T. Clark. 1999.

Moltmann, Jürgen, and Elisabeth Moltmann-Wendel. *God—His and Hers*. Translated by John Bowden. New York: Crossroad, 1991.

Moltmann, Jürgen, and Hans Küng. *Conflicts about the Holy Spirit*. New York: Seabury, 1979.

Moltmann, Jürgen, and Johannes Metz, editors. *Faith and the Future: Essays on Theology, Solidarity and Modernity*. Maryknoll, NY: Orbis, 1995.

Moltmann, Jürgen, and Pinchas Lapide. *Jewish Monotheism and Christian Trinitarian Doctrine*. Translated by Leonard Swidler. Philadelphia: Fortress, 1981.

SECONDARY SOURCES

Ansell, Nicholas John. "The Annihilation of Hell: Universal Salvation and the Redemption of Time in the Eschatology of Jürgen Moltmann." PhD diss., Vrije Universiteit, 2005.

Aulen, Gustav. *Christus Victor: An Historical Study of the Three Main Types of the Idea of the Atonement.* London: SPCK, 1961.

Aung, Salai Hla. *The Doctrine of Creation in the Theology of Barth, Moltmann and Pannenberg: Creation in Theological, Ecological and Philisophical Scientific Perspective.* Regensburg: Roderer, 1998.

Baillie, Donald M. *The Theology of the Sacraments and other Papers.* New York: Scribners, 1957.

Barbour, Ian G. *Nature, Human Nature, and God.* Minneapolis: Fortress, 2002.

———. *When Science Meets Religion: Enemies, Strangers, or Partners?* San Francisco: Harper San Francisco, 2000.

Barth, Karl. *Church Dogmatics* I/1; II/2; IV/1. Translated by G.W. Bromiley. Edinburgh: T. & T. Clark, 1975.

Basinger, David. "Divine Persuasion: Could the Process God Do More?" *Journal of Religion* 64:3 (1984) 332–47.

Bauckham, Richard. *God Will Be All in All: The Eschatology of Jürgen Moltmann.* Edinburgh: T. & T. Clark, 1999.

———. "Jesus and the Wild Animals (Mark 1:13) A Christological Image for an Ecological Age." In *Jesus of Nazareth Lord and Christ: Essays on the Historical Jesus and the New Testament Christology,* edited by Joel B. Green and Max Turner. Grand Rapids: Eerdmans, 1994.

———, editor. *Moltmann: Messianic Theology in the Making.* Basingstroke: Marshall Pickering, 1987.

———. "Moltmann's Theology of Hope Revisited." *Scottish Journal of Theology* 42:2 (1989) 199–214.

———. *The Theology of Jürgen Moltmann.* Edinburgh: T. & T. Clark, 1995.

Becker, William H. "Ecological Sin." *Theology Today* 49:2 (1992) 152–64.

Beek, Huibert Van. Edited by *Sharing Life.* Geneva: World Council of Churches, 1989.

Berry, Thomas. *The Dream of the Earth.* San Francisco: Sierra Club, 1988.

———. *The Great Work: Our Way into the Future.* New York: Bell Tower, 1999.

———. "The Catholic Church and the Religions of the World." *Riverdale Papers* X, (1985) 1–15.

Berry, Thomas, and Brian Swimme. *The Universe Story.* San Francisco: Harper, 1992.

Berry, Thomas, and Thomas Clarke. *Befriending the Earth: A Theology of Reconciliation between Humans and the Earth.* Edited by Stephen Dunn and Anne Lonergan. Mystic, CT: Twenty-Third, 1991.

Berry, Wanda Warren. "Images of Sin and Salvation in Feminist Theology." *Anglican Theological Review* 60:1 (1978) 25–54.

Best, Thomas F., and Dagmar Heller, editors. *So We Believe, So We Pray: Towards Koinonia in Worship.* Faith and Order Paper 171. Geneva: World Council of Churches, 1995.

Best, Thomas F., and Gunther Gassmann, editors. *On the Way to Fuller Koinonia.* Faith and Order Paper 166. Geneva: World Council of Churches, 1994.

Best, Thomas F., and Wesley Granberg-Michaelson, editors. *Koinonia and Justice, Peace and Creation: Costly Unity.* Translated by Geneva: World Council of Churches, 1993.

Boff, Leonardo. *Cry of the Earth, Cry of the Poor.* Maryknoll, NY: Orbis, 1997.
———. *Ecology and Liberation: A New Paradigm.* Maryknoll, NY: Orbis, 1997.
———. *Jesus Christ Liberator: A Critical Christology for Our Time.* Maryknoll, NY: Orbis, 1978.
———. *Passion of the Christ, Passion of the World: The Facts, Their Interpretation, and Their Meaning Yesterday and Today.* Maryknoll, NY: Orbis, 1987.
———. *Trinity and Society.* Maryknoll, NY: Orbis, 1988.
Boff, Leonardo, and Virgil Elizondo, editors. *Ecology and Poverty: Cry of the Earth, Cry of the Poor.* Maryknoll, NY: Orbis, 1995.
Bosch, David, *Transforming Mission: Paradigm Shifts in Theology of Mission.* Maryknoll, NY: Orbis, 1991.
Bouma-Prediger, Steven. "Creation As the Home of God: The Doctrine of Creation in the Theology of Jürgen Moltmann." *Calvin Theological Journal* 32 (1997) 72–90.
———. *The Greening of Theology: The Ecological Models of Rosemary Radford Ruether, Joseph Sitlter, and Jürgen Moltmann.* Atlanta: Scholars, 1995.
Bouma-Prediger, Steven, and Peter Bakke, editors. *Evocations of Grace: The Writings of Joseph Sittler on Ecology, Theology, and Ethics.* Grand Rapids: Eerdmans, 2000.
Bracken, Joseph. "Process Philosophy and Trinitarian Theology," (2 parts) *Process Studies* 8 (1978); 217–30; (1981) 83–96.
Bratton, Susan Power. "Ecofeminism and the Problem of Divine Immanence: Transcendence." *Science & Christian Belief* 6:1 (1994) 21–40.
Breck, John. "The Lord is the Spirit: An Essay in Christological Pneumatology." *Ecumenical Review* 42:2 (1990) 114–21.
Brock, Rita Nakashima. *Journey by Heart: A Christology of Erotic Power.* New York: Crossroad, 1988.
Brown, John P. "The Holy Spirit in the Struggles of People for Liberation and Fullness of Life." *International Review of Mission* 79 (1990) 273–81.
Brueggemann, Walter. *Genesis.* Atlanta: John Knox, 1982.
Bryant, David J. "*Imago Dei,* Imagination, and Ecological Responsibility." *Theology Today* 57:1 (2000) 35–50.
Bryant, M. Darrol. "The Modern Myth of Mastery and the Christian Doctrine of Creation: A Journey in Ecology and Creation Theology." *Dialogue and Alliance* 9:2 (1995) 56–68.
Burston, Daniel. *Erik Erikson and the American Psyche: Ego, Ethics, and Evolution.* Lanham, MD: Rowman & Littlefield, 2007.
Buxton, Graham, *The Trinity, Creation and Pastoral Ministry: Imaging the Perichoretic God.* Wanesboro, GA: Paternoster, 2005.
Carpenter, James A. *Nature and Grace: Toward an Integral Perspective.* New York: Crossroad, 1988.
Carr, Anne. *Transforming Grace: Christian Tradition and Women's Experience.* New York: Harper & Row, 1990.
Carter, Dee. "Foregrounding the Environment: The Redemption of Nature and Jürgen Moltmann's Theology." *Ecotheology* 10 (2001) 70–84.
Castleman, Robbie. "The Evangelical Spirituality of Creation Care and the Kingdom of God." In *For All the Saints: Evangelical Theology and Christian Spirituality,* edited by Timothy George and Alister McGrath, 155–64. Louisville: Westminster John Knox, 2003.

Chapple, Christopher Key, and Mary Evelyn Tucker editors. *Hinduism and Ecology: the Intersection of Earth, Sky, and Water.* Cambridge: Harvard University Press, 2000.

Cho, Hyun-Chul. *An Ecological Vision of the World: Toward a Christian Ecological Theology for Our Age.* Rome: Gregorian University Press, 2004.

Christiansen, Drew, and Walter Grazer, editors. *And God Saw that It Was Good.* Washington, DC: United States Catholic Conference, 1996.

Coakley, Sarah. "What Does Chalcedon Solve and What Does it Not? Some Reflections on the Status and Meaning of Chalcedonian 'Definition.'" In *The Incarnation: An Interdisciplinary Symposium on the Incarnation of the Son of God*, edited by Stephen T. Davis, Daniel Kenall, and Gerald O'Collins, 140–63. New York: Oxford University Press, 2002.

Cobb, John B. Jr. *Christ in a Pluralistic Age.* Philadelphia: Westminster, 1975.

———. *A Christian Natural Theology Based on the Thought of Alfred North Whitehead.*

———. "Christian Theism and The Ecological Crisis." *Religious Education* 66 (1971).

———. "Ecology, Science, and Religion: Toward A Postmodern Worldview." In *The Reenchantment of Science: Postmodern Proposals*, edited by David R. Griffin. Albany: State University of New York Press, 1988.

———. *God and the World.* Philadelphia: Westminster, 1965.

———. *Is it Too Late? A Theology of Ecology.* Beverly Hills, CA: Bruce, 1972. Philadelphia: Westminster, 1974.

———. "Jürgen Moltmann's Ecological Theology in Process Perspective." *Asbury Theological Journal* 55:1 (2000) 115–28.

———. "Process Theology and Environmental Issues." *Journal of Religion* 60:4 (1980) 440–58.

———. "Process Theology as an Ecological Model." In *Cry of the Environment*, edited by Philip N. Joranson and Ken Butigan. Santa Fe: Bear, 1984.

———. "Reply to Jürgen Moltmann's 'The Unity of the Triune God.'" *St Vladimir's Theological Quarterly* 28 (1984) 173–77.

Cobb, John B. Jr., and David Ray Griffin. *Process Theology: An Introductory Exposition.* Philadelphia: Westminster, 1976.

Cobb, John B. Jr., and Clark H. Pinnock, editors. *Searching for an Adequate God: A Dialogue between Process and Free Will Theists.* Grand Rapids: Eerdmans, 2000.

Cochrane, Arthur C. *The Church's Confession under Hitler.* Philadelphia: Westminster, 1962.

Coffey, David. "Spirit Christology and the Trinity." In *Advents of the Spirit: An Introduction to the Current Study of Pneumatology*, edited by Bradford E. Hinze & D. Lyle Dabney. Milwaukee: Marquette University Press, 2001.

Coles, Robert. *Erik H. Erikson: The Growth of His Work.* Boston: Little Brown, 1970.

Comblin, Jose. *The Holy Spirit and Liberation.* Maryknoll, NY: Orbis, 1989.

Conradie, Ernst M. *The Church and Climate Change.* Signs of the Times 1. Pietermaritzburg: Cluster, 2008.

———. "Towards an Ecological Reformulation of the Christian Doctrine of Sin." *Journal of Theology for Southern Africa* 122 (2005) 4–22.

Conyers, A. J. *God, Hope, and History: Jürgen Moltmann and the Christian Concept of History.* Macon, GA: Mercer University Press, 1988.

Cooper, John W. *Body, Soul & Life Everlasting: Biblical Anthropology and the Monism-Dualism Debate.* Grand Rapids: Eerdmans, 1989.

Dabney, D. Lyle. "*Pneumatologia Crucis:* Reclaiming *Theologia Crucis* for a Theology of the Spirit Today." *Scottish Journal of Theology* 53:4 (2000) 511–24.

Davidson, Mark. *Uncommon Sense: The Life and Thought of Ludwig von Bertalanffy, Father of General Systems Theory.* Los Angeles: J. P. Tarchers, 1983.

Davis, John Jefferson. "Ecological 'Blind Spot' in the Structure and Content of Recent Evangelical Systematic Theologies." *Journal of the Evangelical Theological Society* 43:2 (2000) 273–86.

Deane-Drummond, Celia E. *Creation Through Wisdom: Theology and the New Biology.* Edinburgh: T. & T. Clark, 2000.

———. *Ecology in Jürgen Moltmann's Theology.* Lewiston, NY: Edwin Mellen, 1997.

Diamond, Irene, and Gloria Femen Orenstein, editors. *Reweaving the World: The Emergence of Ecofeminism.* San Francisco: Sierra Club, 1990.

Duby, Georges. *The Age of the Cathedrals.* Chicago: University of Chicago Press, 1981.

Duffy, Stephen J. "Our Hearts of Darkness: Original Sin Revisited." *Theological Studies* 49 (1988) 597–622.

Dulles, Avery Robert. "The Population of Hell." *First Things* 133:1 (2003) 36–41.

Edwards, Denis. *Breath of Life: A Theology of the Creator Spirit.* Maryknoll, NY: Orbis, 2004.

———, editor. *Earth Revealing-Earth Healing: Ecology and Christian Theology,.* Collegeville, MN: Liturgical, 2001.

———. "Ecology and the Holy Spirit: The "Already" and the "Not Yet" of the Spirit in Creation." *Pacifica* 13 (2000) 142–59.

———. *The God of Evolution: A Trinitarian Theology.* New York: Paulist, 1999.

———. *Jesus and the Cosmos.* New York: Paulist, 1991.

———. *Jesus the Wisdom of God: An Ecological Theology.* Maryknoll, NY: Orbis, 1995.

Elsbernd, Mary. "Toward a Theology of Spirit That Builds Up the Just Community." In *The Spirit in the Church and the World,* edited by Bradford E. Hinze. Maryknoll, NY: Orbis, 2004.

Fagerberg, David W. *Theologia Prima: What is Liturgical Theology?* Chicago: Hillenbrand, 2004.

Faith and Order. *Fifth World Conference in Faith and Order: Message Section Reports Discussion Paper.* Faith and Order Paper:164. Geneva: World Council of Churches, 1993.

———. *Towards Koinonia in Faith, Life and Witness: A Discussion Paper.* Faith and Order Paper 161. Geneva: World Council of Churches, 1993.

Farrow, Doublas B. "Review Essay: In the End is the Beginning: A Review of Jürgen Moltmann's Systematic Contributions." *Modern Theology* 14:3 (1998) 425–47.

Fensham, Charles. *Emerging from the Dark Age Ahead: The Future of the North American Church.* Toronto: Novalis, 2008.

Ferguson, Sinclair B., and David F. Wright, editor. *New Dictionary of Theology.* Downers Grove, IL: InterVarsity, 1988.

Finger, Thomas. "Trinity, Ecology and Panentheism." *Christian Scholar's Review* 27:1 (1997) 74–98.

Foley, Edward, Kathleen Hughes, and Gilbert Ostdiek. "The Preparatory Rites: A Case Study in Liturgical Ecology." In *The Ecological Challenge: Ethical, Liturgical, and Spiritual Responses,* edited by Richard N. Fragomeni and John T. Pawlikowski, 67–101. Collegeville, MN: Liturgical, 1994.

Fox, Matthew. *The Coming of the Cosmic Christ: The Healing of Mother Earth and the Birth of a Global Renaissance*. San Francisco: Harper & Row, 1988.

———. "Creation-Centered Spirituality from Hildegard of Bingen to Julian of Norwich: 300 Years of an Ecological Spirituality in the West." In *Cry of the Environment: Rebuilding the Christian Creation Tradition*, edited by Philip N. Joranson and Ken Butigan, 85–106. Santa Fe, NM: Bear, 1984.

———. *Original Blessing*. Santa Fe, NM: Bear, 1983.

Fragomeni, Richard N. "Liturgy at the Heart of Creation: Towards an Ecological Consciousness in Prayer." In *The Ecological Challenge: Ethical, Liturgical, and Spiritual Responses*, edited by Richard N. Fragomeni and John T. Pawlikowski. Collegeville, MN: Liturgical, 1994.

French, William C. "Returning to Creation: Moltmann's Eschatology Naturalized." *The Journal of Religion* 68:1 (1988) 78–86.

Fretheim, Terence E. *The Suffering of God: An Old Testament Perspective*. Philadelphia: Fortress, 1984.

Galloway, Allan D. *The Cosmic Christ: A Development and Exposition of the Doctrine of Cosmic Redemption in Biblical Theology*. New York: Harper, 1951.

Gamble, Richard C., editor. *Calvin's Ecclesiology: Sacrmaents and Deacons*. New York: Garland, 1992.

Gerrish, B. A., editor. *Reformed Theology for the Third Christian Millennium*. Louisville: Westminster John Knox, 2003.

Gilkey, Langdon. *Maker of Heaven and Earth*. New York: Doubleday, 1959.

———. *Nature, Reality and the Sacred: The Nexus of Science and Religion*. Minneapolis: Fortress, 1993.

Goetz, Ronald. "The Suffering God: The Rise of a New Orthodoxy." *The Christian Century*, April 16, 1986, 385–89.

Grenz, Stanley J. *The Social God and the Relational Self: A Trinitarian Theology of the Imago Dei*. Louisville: Westminister John Knox, 2001.

Griffin, David Ray. "*Creatio Ex Nihilo*, The Divine *Modus Operandi*, and The *Imitatio Dei*." In *Faith and Creativity: Essays in Honor of Eugene H. Peters*, edited by George Nordgulen and George W. Shields. St. Louis: CBP, 1987.

———. "Creation out of Chaos and the Problem of Evil." In *Encountering Evil: Live Options in Theodicy*, edited by Stepthen T. Davis. Atlanta: John Knox, 1981.

———. *Evil Revisited: Responses and Reconsiderations*. Albany: SUNY Press, 1991.

———. *God, Power, and Evil: A Process Theodicy*. Philadelphia, Westminster: 1976.

Grim, John A., editor. *Indigenous Traditions and Ecology: the Interbeing of Cosmology and Community*. Cambridge: Harvard Univerity Press, 2001.

Grim, John A. "Living in a Universe: Native Cosmologies and the Environment." In *When Worlds Converge: What Science and Religion Tell Us about the Story of the Universe and Our Place in It*, edited by Clifford N. Matthews, Mary Evelyn Tucker, and Philip Hefner, 243–60. Chicago: Open Court, 2002.

Gulick, Walter B. "The Bible and Ecological Spirituality." *Theology Today* 48 (1991) 182–94.

Gunton, Colin. *The Triune Creator: A Historical and Systematic Study*. Edinburgh: Edinburgh University Press, 1988.

Haight, John F. *God After Darwin: A Theology of Evolution*. Boulder, CO: Westview, 2000.

Haight, Roger, "The Case for Spirit Christology." *Theological Studies* 53 (1992) 257–87.

Hall, Douglas John. *Imaging God: Dominion as Stewardship*. Grand Rapids: Eerdmans, 1986.
Harrison, Verna. "Perichoresis in the Greek Fathers." *St. Vladimir's Theological Quarterly* 35:1 (1991) 53–65.
Hart, Trevor. "A Capacity for Ambiguity? The Barth-Brunner Debate Revisited." *Tyndale Bulletin* 44:2 (1993).
Hartshorne, Charles. *Man's Vision of God and the Logic of Theism*. Hamden, CT: Archon, 1964.
———. *A Natural Theology for Our Time*. Chicago: Open Court, 1967.
———. "A New Look at the Problem of Evil." In *Current Philosophical Issues: Essays in Honor of Curt John Ducasse*, edited by F. C. Dommeyer. Springfield, IL: Thomas, 1966.
Harvey, John A. *A Handbook of Theological Terms*. New York: Macmillan, 1964.
Harvey, Van A. *A Handbook of Theological Terms*. New York: Macmillan, 1964.
Hastings, James, editor. *Encyclopedia of Religion and Ethics*. New York: Scribners, 1911.
Hendry, George S. *Theology of Nature*. Philadelphia: Westminster, 1980.
Hessel, Dieter T., and Rosemary Radford Reuther, editors. *Christianity and Ecology: Seeking the Well-Being of Earth and Humans*. Cambridge: Harvard University Press, 2000.
Hiebert, Theodore. "The Human Vocation: Origins and Transformations in Christian Traditions." In *Christianity and Ecology: Seeking The Well-Being of Earth and Humans*, edited by Dieter T. Hessel and Rosemary Radford Ruether, 135–51. Cambridge: Harvard University Press, 2000.
———. "Rethinking Traditional Approaches to Nature in the Bible." In *Theology for Earth Community: A Field Guide*, edited by Dieter T. Hessel. Maryknoll, NY: Orbis, 1996.
Hiers, Richard H. "Ecology, Biblical Theology, and Methodology: Biblical Perspectives on the Environment." *Zygon* 19:1 (1984)
Honner, John, "A New Ontology: Incarnation, Eucharist, Resurrection, and Physics." *Pacifica* 4 (1991) 15–50.
Hunsinger, George. "The Crucified God and the Political Theology of Violence: A Critical Survey of Jürgen Moltmann's Recent Thought (I)." *Heythrop Journal* 14 (1973) 270–79.
Jacobson, Diane. "Biblical Bases for Eco-Justice Ethics." In *Theology for Earth Community: A Field Guide*, edited by Dieter T. Hessel. Maryknoll, NY: Orbis, 1996.
Jansen, Henry. "Moltmann's View of God's (Im)mutability: The God of the Philosophers the God of the Bible." *Neue Zeitschrift für Systematische Theologie und Religionsphilosophie* 36:3 (1994) 284–301.
Jantzen, Grace. *God's World, God's Body*. Philadelphia: Westminster, 1984.
Jenson, Robert W. "Justification as a Triune Event." *Modern Theology* 11:4 (1995) 421–28.
Jeremias, Joachim. *The Eucharistic Words of Jesus*. London: SCM, 1966.
John, V. J., "Ecology in the Parables: The Use of Nature Language in the Parables of the Synoptic Gospel." *Asia Journal of Theology* 14:2 (2000) 304–27.
Johnson, Elizabeth A. *Friends of God and Prophets: A Feminist Theological Reading of the Communion of Saints*. Continuum: New York, 1999.
———. *She Who Is: The Mystery of God in Feminist Theological Discourse*. New York: Crossroad, 1992.

Kärkkäinen, Veli-Matti. "The Holy Spirit and Justification: The Ecumenical Significance of Luther's Doctrine of Salvation." *PNEUMA: The Journal of the Society for Pentecostal Studies* 24:1 (2002) 26–39.

———. *The Trinity: Global Perspectives*. Louisville: Westminster John Knox, 2007.

Kearney, Richard. *The God Who May Be: A Hermenutics of Religion*. Bloomington: Indiana University Press, 2001.

———. *On Stories: Thinking in Action*. London: Routledge, 2002.

———. *The Wake of Imagination: Toward a Postmodern Culture*. Minneapolis: University of Minnesota Press, 1988.

Kearns, Laurel, and Catherine Keller, editors. *Ecospirit: Religion and Philosophies for the Earth*. New York: Fordham University Press, 2007.

Keller, Catherine. "Pneumatic Nudges: The Theology of Moltmann, Feminism, and the Future." In *The Future of Theology: Essays in Honor of Jürgen Moltmann*, edited by Miroslav Volf, Carmen Krieg, and Thomas Kucharz. Grand Rapids: Eerdmans, 1996.

Kenel, Sally. "Nature and Grace: An Ecological Metaphor." In *An Ecology of the Spirit: Religious Reflection and Environmental Consciousness*, edited by Michael Barnes. New York: University Press of America, 1994.

Kistler, Don. Edited by *Sola Scriptura!* Morgan, PA: Soli Deo Gloria, 1995.

Klug, Lyn, editor. *All Will Be Well: A Gathering of Healing Prayer*. Minneapolis: Fortress, 1998.

Komonchak, Joseph A., Mary Collins, and Dermot A. Lane, editors. *The New Dictionary of Theology*. Wilmington, DE: Michael Glazier, 1988.

Koyama, Kosuke. *Mount Fuji and Mount Sinai: A Pilgrimage in Theology*. London: SCM, 1984.

Knitter, Paul F. "Deep Ecumenicity versus Incommensurability: Finding Common Ground on a Common Earth." In *Christianity and Ecology: Seeking the Well-Being of Earth and Humans*, edited by Dieter T. Hessel and Rosemary Radford Ruether, 374–76. Cambridge: Harvard University Press, 2000.

———. "A New Pentecost? A Pneumatological Theology of Religions." *Current Dialogue* 19 (1991) 32–41.

———. "The Relational God: Aquinas and Beyond." *Theological Studies* 46 (1985) 647–63.

Korean Hymnal Society. *Korean-English Hymnal*. Seoul, Korea: Christian Literature Society of Korea, 1985.

Koyama, Kosuke. *Mount Fuji and Mount Sinai: A Pilgrimage in Theology*. London: SCM, 1984.

———. *Waterbuffalo Theology*. London: SCM, 1974.

Lai, Pan-chiu. "Christian Ecological Theology in Dialogue with Confucianism." *Ching Feng* 41:3–4 (1998) 309–35.

Lapide, Pinchas. "Jewish Monotheism." In *Jewish Monotheism and Christian Trinitarian Doctrine: A Dialogue in Pinchas Lapide & Jürgen Moltmann*. Philadelphia: Fortress, 1979.

Larkin, Ernest E. "Asceticism." In *The New Dictionary of Theology*, edited by Joseph A. Komonchak, Mary Collins, and Dermot A. Lane. Wilmington, DE: Michael Glazier, 1987.

Lathrop, Gordon W. *Holy Ground: A Liturgical Cosmology*. Minneapolis: Augsburg Fortress, 2003.

———. *Holy People: A Liturgical Ecclesiology*. Minneapolis: Fortress, 1999.

———. *Holy Things: A Liturgical Theology*. Minneapolis: Fortress, 1993.

Lee, Bryan Jeong Guk. "An Eschatological Apostolic Understanding of the Authority of the Bible." ThM thesis, Knox College, Toronto School of Theology, 2005.

Lee, Jai-Don. "Towards an Asian Ecotheology in the Context of Thomas Berry's Cosmology: a Critical Inquiry." ThD diss., University of St. Michael's College, Toronto School of Theology, 2004.

Lee, Jeong-Woo. "Toward a Trinitarian Ecological Theology: a Study in Jürgen Moltmann's Panentheism." PhD diss., University of St. Michael's College, Toronto, 2007.

Lee, Jung Young. *God Suffers for Us: A Systematic Inquiry into a Concept of Divine Passibility*. Hague: Martinus Nijhoff, 1974.

———. *The Trinity in Asian Perspective*. Nashville: Abingdon, 1996.

Leech, Kenneth. *The Social God*. London: Sheldon, 1981.

Linahan, Jane E. "Experiencing God in Brokenness: The Self-Emptying of the Holy Spirit in Moltmann's Pneumatology." In *Encountering Transcendence: Contributions to A Theology of Christian Religious Experience*, edited by Lieven Boeve, Hans Geybels, and Stijn Van den Bossche, 165–84. Leuven: Peeters, 2005.

———. "The Grieving Spirit: The Holy Spirit as Bearer of the Suffering of the World in Moltmann's Pneumatology." In *The Spirit in the Church and the World*, edited by Bardford E Hize. Maryknoll, NY: Orbis, 2004.

Linzey, Andrew. "Unfinished Creation; The Moral and Theological Significance of the Fall." *Ecotheology* 4 (1998) 20–26.

Lonergan, Anne, and Caroline Richards editors. *Thomas Berry and the New Cosmology*. Mystic, CT: Twenty-Third, 1987.

Lovejoy, Arthur O. *The Great Chain of Being: A Study of the History of an Idea*. Cambridge: Harvard University Press, 1942.

Macchia, Frank D. "Justification through New Creation: The Holy Spirit and the Doctrine by Which the Church Stands or Falls." *Theology Today* 58:2 (2001) 202–17.

Maimela, Simon S. "The Atonement in the Context of Liberation Theology." *International Review of Mission* 75 (1986) 261–69.

Mangina, Joseph L. *Karl Barth: Theologian of Christian Witness*. Louisville: Westminster John Knox, 2004.

Mannion, Gerard, and Lewis S. Mudge, editor. *The Routledge Companion to the Christian Church*. New York: Routledge, 2008.

Maxwell, Jack Martin. *Worship and Reformed Theology: The Liturgical Lessons of Mercersburg*. Pittsburgh, PA: Pickwick, 1976.

McAffe, Gene. "Ecology and Biblical Studies." In *Theology for Earth Community: A Field Guide*, edited by Dieter T. Hessel. Maryknoll, NY: Orbis, 1996.

McCarthy, Scott. *Celebrating the Earth*. San Jose, CA: Resource, 1991.

McDaniel, Jay B. *Earth, Sky, Gods and Mortals: Developing an Ecological Spirituality*. Mystic, CT: Twenty-Third, 1990.

McDougall, Joy Ann. "The Return of Trinitarian Praxis? Moltmann on the Trinity and the Christian Life." *Journal of Religion* 83:2 (2003) 177–203.

McFague, Sallie. *The Body of God: An Ecological Theology*. Minneapolis: Fortress, 1993.

———. "An Ecological Christology: Does Christianity Have It?" In *Christianity and Ecology: Seeking The Well-Being of Earth and Humans*, edited by Dieter T. Hessel and Rosemary Radford Ruether. Cambridge: Harvard University Center for the Study of World Religions, 2000.

———. *Models of God: Theology for an Ecological, Nuclear Age*. Philadelphia: Fortress, 1987.

———. *Super, Natural Christians: How We Should Love Nature*. Minneapolis: Fortress, 1997.

McPherson, Jim. "The Integrity of Creation: Science, History and Theology." *Pacifica* 2 (1989) 333–55.

McWilliams, Warren. "Christic Paradigm and Cosmic Christ: Ecological Christology in the Theologies of Sallie McFague and Jürgen Moltmann." *Perspectives in Religious Studies* 25.4 (1998) 341–55.

———. "Divine Suffering in Contemporary Theology." *Scottish Journal of Theology* 33 (1980) 35–53.

———. *The Passion of God: Divine Suffering in Contemporary Theology*. Macon, GA: Mercer University Press, 1985.

———. "Spirit Christology and Inclusivism: Clark Pinnock's Evangelical Theology of Religions." *Perspectives in Religious Studies* 24:3 (1997) 325–36.

Meland, Bernard Eugene. "New Perspectives on Nature and Grace." In *The Scope of Grace: Essays on Nature and Grace in Honor of Joseph Sittler*, edited by Philip J. Hefner. Philadelphia: Fortress, 1964.

Metropolitan Gennadios of Sassima, editor. *Grace in Abundance: Orthodox Reflections on the Way to Porto Alegre*. Geneva: World Council of Churches, 2005.

Mick, Lawrence E. "The Ecology of Worship." In *Liturgy and Ecology in Dialogue*. Collegeville, MN: Liturgical, 1997.

Middleton, J. Richard. "The Liberating Image? Interpreting the *Imago Dei* in Context." *Christian Scholar's Review* 24:1 (1994) 8–25.

Min, Anselm Kyongsuk. "Liberation, the Other, and Hegel in Recent Pneumatologies." *Religious Studies Review* 22:1 (1996) 28–33.

———. "Solidarity of Others in the Power of the Holy Spirit: Pneumatology in a Divided World." In *Advents of the Spirit: An Introduction to the Current Study of Pneumatology*, edited by Bradford E. Hinze and D. Lyle Dabney. Milwaukee: Marquette University Press, 2001.

Molnar, Paul D. "The Function of the Trinity in Moltmann's Ecological Doctrine of Creation." *Theological Studies* 51 (1990) 673–97.

Mozley, J. K. B. D. *The Impassibility of God: A Survey of Christian Thought*. London: Cambridge University Press, 1926.

Muller, Richard A. *Christ and the Decree: Christology and Predestination in Reformed Theology from Calvin to Perkins*. Grand Rapids: Baker, 1986.

Nash, James. *Loving Nature: Ecological Integrity and Christian Responsibility*. Nashville: Abingdon, 1991.

Neville, Robert Cummings. *Creativity and God: A Challenge to Process Theology*. New York: SUNY, 1995.

———. *God the Creator: On the Transcendence and the Presence of God*. Chicago: University of Chicago Press, 1986.

Nolan, Barbara. *The Gothic Visionary Perspective*. Princeton, NJ: Princeton University Press, 1977.

Nnamani, Amuluche Gregory. *The Paradox of a Suffering God: On the Classical, Modern-Western and Third World Struggles to Harmonize the Incompatible Attributes of the Trinitarian God*. Frankfurt am Main: Peter Lang, 1995.

O'Donnell, John, *Trinity and Temporality: The Christian Doctrine of God in light of Process Theology and the Theology of Hope.* New York: Oxford University Press, 1983.

———. "The Trinity as Divine Community: A Critical Reflection upon Recent Theological Developments." *Gregorianum* 69:1 (1988) 5–34.

O'Keefe, Mark. *What Are They Saying About Social Sin?* New York: Paulist, 1990.

O'Neill, Mark. *Cathedrals.* London, England: Roydon, 1984.

Old, Hughes Oliphant. *Worship: Reformed According to Scripture.* Louisville: Westminster John Knox, 2002.

Olsen, R. "Trinity and Eschatology: The Historical Being of God in Jürgen Moltmann and Wolfhard Pannenberg." *Scottish Journal of Theology* 36 (1983) 213–27.

Otto, Randall E. "The Use and Abuse of Perichoresis." *Scottish Journal of Theology* 54 (2001) 366–84.

Page, Ruth. "The Influence of the Bible on Christian Belief about the Natural World." In *Christianity and Ecology*, edited by Elizabeth Breuilly and Martin Palmer. New York: Cassell, 1992.

Pailin, David A. "The Incarnation as a Continuing Reality." *Religious Studies* 6:4 (1970) 303–27.

Panofsky, Erwin. *Gothic Architecture and Scholasticism.* New York: Meridian, 1957.

Park, Mann. "Jürgen Moltmann's Theology of the Trinity and Its Significance for Contemporary Social Questions: A Dialogical Approach." PhD diss., University of St. Michael's College, Toronto School of Theology, 2000.

Park, Seong-Won. "Worship in the Presbyterian Church in Korea." In *Christian Worship in Reformed Churches Past and Present*, edited by Lukas Vischer. Grand Rapids: Eerdmans, 2003.

Parsons, Susan Frank, editor. *The Cambridge Companion to Feminist Theology.* Cambridge: Cambridge University Press, 2002.

Pearson, Andrew. *Making Creation Visible: God's Earth in Christian Worship.* Cambridge: Grove, 1996.

Persaud, Winston D. "Toward a Trinitarian Theology of Justification and Vision of Eco-Justice." *Dialog* 31 (1992) 294–302.

Peter, Scott. "The Future of Creation: Ecology and Eschatology." In *The Future as God's Gift: Explorations in Christian Eschatology*, edited by David Fergusson and Marcel Sarot. Edinburgh: T. & T. Clark, 2000.

Peters, Ted, Robert John Russell, and Michael Welker, editors. *Resurrection: Theological and Scientific Assessments.* Grand Rapids: Eerdmans, 2002.

Peterson, Gregory R. "Whither Panentheism?" *Zygon* 36:3 (2001) 395–405.

Polkinghorne, John. *The Faith of A Physicist: Reflections of a Bottom-Up Thinker.* Minneapolis: Fortress, 1996.

The Presbyterian Church in Canada, Task Force on the Revision of the Book of Praise. *The Book of Praise.* Toronto, Ontario: The Presbyterian Church in Canada, 1997.

The Presbyterian Church in Canada, The Worship Committee, Board of Congregational Life. E. Margaret MacNaughton, editor. *The Book of Common Worship.* Toronto, Ontario: The Presbyterian Church in Canada, 1991.

Rasmussen, Larry. *Earth Community, Earth Ethics.* Maryknoll, NY: Orbis, 1996.

Rayan, Samul, "The Earth is the Lord's." In *Ecotheology: Voices from South and North*, edited by David G Hallman. Maryknoll, NY: Orbis, 1994.

Regidor, Jose Ramos. "Some Premises for an Eco-Social Theology of Liberation." In *Ecology and Poverty: Cry of the Earth, Cry of the Poor*, edited by Leonardo Boff and Virgil Elizondo, 78–93. London: SCM, 1995.

Rice, Howard, and James C. Huffstutler. *Reformed Worship*. Louisville: Geneva, 2001.

Richardson, Alan. *A Dictionary of Christian Theology*. London: SCM, 1969.

Rolston, Holmes III. "Does Nature Need To Be Redeemed?" *Zygon* 29:2 (1994) 205–29.

Romano, Eugene L. *A Way of Desert Spirituality: The Plan of Life of the Hermits of Bethlehem*. Chester, NJ: Hermits of Bethlehem and the Heart of Jesus, 1998.

Rosato, Philip J., SJ. "Spirit Christology: Ambiguity and Promise." *Theological Studies* 38 (1977) 423–49.

Ross, Robert H., editor. *Alfred, Lord Tennyson In Memoriam: An Authoritative Text Backgrounds and Sources Criticism*. New York: Norton, 1973.

Ruether, Rosemary R. "Ecofeminism: Symbolic and Social Connections of the Oppression of Women and the Domination of Nature." In *Ecofeminism and the Sacred*, edited by Carol J. Adams. New York: Continuum, 1993.

———. "The First and Final Proletariat: Socialism and Women's Liberation." *Soundings* 58 (1975) 310–27.

———. *Gaia and God: An Ecofeminist Theology of Earth Healing*. San Francisco: Harper San Francisco, 1992.

———. *Liberation Theology: Human Hope Confronts Christian History and American Power*. NY: Paulist, 1972.

———. *Sexism and God-Talk: Toward a Feminist Theology*. Boston: Beacon, 1983.

———, editor. *Women Healing Earth: Third World Women on Ecology, Feminism and Religion*. Maryknoll, NY: Orbis, 1996.

Russell, Robert John. "Finite Creation without a Beginning: The Doctrine of Creation in Relation to Big Bang and Quantum Cosmologies." In *Quantum Cosmology and the Laws of Nature: Scientific Perspectives on Divine Action*, edited by Robert John Russell, Jancy C. Murphy, and Chris J. Isham. Berkeley, CA: Center for Theology and the Natural Sciences, 1993.

Saliers, Don E. *Worship as Theology: Foretaste of Glory Divine*. Nashville: Abingdon, 1994.

Santmire, H. Paul. *Nature Reborn: The Ecological and Cosmic Promise of Christian Theology*. Minneapolis: Fortress, 2000.

———. *The Travail of Nature: The Ambiguous Ecological Promise of Christian Theology*. Philadelphia: Fortress, 1985.

———. *Ritualizing Nature: Renewing Christian Liturgy in a Time of Crisis*. Minneapolis: Fortress, 2008.

Scharper, Stephen Bede. *Redeeming the Time: A Political Theology of the Environment*. New York: Continuum, 1997.

Schmemann, Alexander. *Introduction to Liturgical Theology*. Portland, ME: The American Orthodox Press, 1966.

Schuurmann, Douglas J. "Creation, Eschaton, and Ethics: An Analysis of Theology and Ethics in Jürgen Moltmann." *Calvin Theological Journal* 22.1 (1987) 42–67.

———. *Creation, Eschaton, and Ethics: The Ethical Significance of the Creation-Eschaton Relation in the Thought of Emil Brunner and Jürgen Moltmann*. New York: Peter Lang, 1991.

Schwarz, Hans. *Creation*. Grand Rapids: Eerdmans, 2002.

Schweitzer, Don. "The Consistency of Jürgen Moltmann's Theology." *Studies in Religion: Sciences Religieuses* 22:2 (1993) 197–208.

———. "Douglas Hall's Critique of Jürgen Moltmann's Eschatology of the Cross." *Studies in Religion* 27:1 (1998) 7–25.

Senior, Donald C. P. "The Earth Story: Where Does the Bible Fit In?" In *Thomas Berry and the New Cosmology*, edited by Anne Lonergan and Caroline Richards. Connecticut: Twenty-Third, 1987.

Senn, Frank C. *Christian Liturgy: Catholic and Evangelical*. Minneapolis: Fortress, 1997.

Simson, Otto Von. *The Gothic Cathedral*. New York: Bollingen Foundation, 1956, 1962.

Sittler, Joseph. *Essays on Nature and Grace*. Philadelphia: Fortress, 1972.

Southgate, Christopher. "God and Evolutionary Evil: Theodicy in the Light of Darwinism." *Zygon* 37:4 (2002) 803–24.

Speidell, Todd Saliba. "The Incarnation As the Hermeneutical Criterion for Liberation and Reconciliation." *Scottish Journal of Theology* 40, (1987) 249–258.

Sproul, R. C. *What is Reformed Theology?* Grand Rapids: Baker, 1997.

Stoeger, William R., "Scientific Accounts of Ultimate Catastrophes in Our Lifebearing Universe." In *The End of the World and the Ends of God: Science and Theology on Eschatology*, edited by John Polkinghorne and Michael Welker. Harrisburg: Trinity, 2002.

Sun, Soon-Hwa. "Human Rights in the Ecological Context." *Reformed World* 48 (1998) 131–40.

Tang, Siu-Kwong. *God's History in the Theology of Jürgen Moltmann*. New York: Peter Lang, 1996.

Tanner, Kathryn. "Eschatology without a Future?" In *The Future As God's Gift: Explorations in Christian Eschatology*, edited by David Fergusson and Marcel Sarot. Edinburgh: T. & T. Clark, 2000.

———. *Jesus, Humanity and the Trinity: A Brief Systematic Theology*. Minneapolis: Fortress, 2001.

Toolan, David. *At Home in the Cosmos*. Maryknoll, NY: Orbis, 2001.

Torrance, Thomas F. *The Christian Doctrine of God: One Being Three Persons*. Edinburgh: T. & T. Clark, 1996.

Toynbee, Arnold. "The Religious Background of the Present Environmental Crisis." In *Ecology and Religion in History*, edited by David and Eileen Spring. New York: Harper & Row, 1974.

Tracy, David. *The Analogical Imagination: Christian Theology and the Culture of Pluralism*. London: SCM, 1981.

———. *Blessed Rage for Order: The New Pluralism in Theology*. New York: Crossroad, 1975.

———. "Theological Method." In *Christian Theology: An Introduction to its Traditions and Task*, edited by Peter C. Hodgson and Robert H. King. Minneapolis: Fortress, 1994.

Tracy, David, and Nicholas Lash, editors. *Cosmology and Theology*. Edinburgh: T. & T. Clark, 1983.

Tubbs, James B., Jr. "Humble Dominion." *Theology Today* 50:4 (1994) 543–56.

Tucker, Mary Evelyn, and Duncan Ryuken Williams. *Buddhism and Ecology*. Cambridge: Harvard University Press, 1997.

Tucker, Mary Evelyn, and John Berthrong, editors. *Confucianism and Ecology*. Cambridge: Harvard University Press, 1998.

United Church of Canada, The. *Celebrate God's Presence: A Book of Services for the United Church of Canada*. Etobicoke, Ontario: United Church, 2000.

―――. *More Voices*. Toronto: United Church, 2007.

―――. *Voices United*. Etobicoke, Ontario: United Church, 1996

Van der Borght, Eddy. "Reformed Ecclesiology." In *The Routledge Companion to the Christian Church*, edited by Gerard Mannion and Lewis S. Mudge. New York: Routledge, 2008.

Van Dyk, Leanne, editor. *A More Profound Alleluia: Theology and Worship in Harmony*. Grand Rapids: Eerdmans, 2005.

Vanhoozer, Kevin J., editor. *The Trinity in a Pluralistic Age: Theological Essays on Culture and Religion*. Grand Rapids: Eerdmans, 1997.

Vanin, Christina. "The Significance of the Incarnation for the Ecological Theology: A Challenging Approach." *Ecotheology* 6:1 (2001) 108–22.

Vischer, Lukas. "The Reformed Tradition and Its Multiple Facets." In The Reformed Family Worldwide: A Survey of Reformed Churches, Theological Schools and International Organizations, edited by Jean-Jacques Bauswein and Lukas Vischer. Grand Rapids: Eerdmans, 1999.

―――."Worship as Christian Witness to Society." In *Christian Worship in Reformed Churches Past and Present*, edited by Lukas Vischer. Grand Rapids: Eerdmans, 2003.

Vischer, Lukas, editor. *Christian Worship in Reformed Churches Past and Present*. Grand Rapids: Eerdmans, 2003.

Vogel, Dwight W., editor. *Primary Sources of Liturgical Theology: A Reader*. Collegeville, MN: Liturgical, 2000.

Volf, Miroslav. *After Our Likeness: The Church as the Image of the Trinity*. Grand Raids: Eerdmans, 1998.

―――. "The Trinity is Our Social Program: The Doctrine of the Trinity and the Shape of Social Engagement." *Modern Theology* 14:3 July (1998) 403–23.

Waite, Willis W., Jr. *Theism, Atheism, and the Doctrine of God: The Trinitarian Theologies of Karl Barth and Jürgen Moltmann in Response to Protest Atheism*. Atlanta: Scholars, 1987.

Wallace, Mark I. *Fragments of the Spirit: Nature, Violence, and the Renewal of Creation*. Harrisburg: Trinity, 2002.

―――. "The Wounded Spirit as the Basis for Hope in an Age of Radical Ecology." In *Christianity and Ecology: Seeking the Well-Being of Earth and Humans*, edited by Dieter T. Hessel and Rosemary Radford Ruether. Cambridge: Harvard University Press, 2002.

Walsh, Brian J. "Theology of Hope and the Doctrine of Creation: An Appraisal of Jürgen Moltmann." *The Evangelical Quarterly* 59:1 (1987).

Walsh, Brian J., Marianne B. Karsh, and Nik Ansell. "Trees, Forestry and the Responsiveness of Creation." In *This Sacred Earth*, edited by Roger S. Gottlieb. New York: Loutledge, 2004.

Wanamaker, Charles. "The Historical Jesus Today: A Reconsideration of the Foundation of Christology." *Journal of Theology for Southern Africa* 94 (1996) 3–17

Warren, Karen J. *Ecofeminism: Women, Culture, Nature*. Bloomington: Indiana University Press, 1997.

―――, editor. *Ecological Feminism*. London: Routledge, 1994.

Welch, Claude. *In This Name: The Doctrine of the Trinity in Contemporary Theology*. New York: Scribners, 1952.

Welker, Michael. *Creation and Reality*. Minneapolis: Fortress, 1999.

———. "The Reign of God." *Theology Today* 49:4 (1993) 500–15.

Wells, Harold G. *The Christic Center: Life-Giving and Liberating*. Maryknoll, NY: Orbis, 2004.

———. "The Flesh of God: Christological Implications for an Ecological Vision of the World." *Toronto Journal of Theology* 15:1 (1999) 51–68.

———. "The Holy Spirit and Theology of the Cross: Significance for Dialogue." *Theological Studies* 53 (1992) 476–81.

———. "Theology of the Cross and the Theologies of Liberation." *Toronto Journal of Theology* 17:1 (2001) 147–66.

Wessels, Cletus, OP. *Jesus in the New Universe Story*. Maryknoll, NY: Orbis, 2003.

White, James F. *Protestant Worship: Traditions in Transition*. Louisville: John Knox, 1989.

White, Lynn, Jr. "The Historical Roots of Our Ecologic Crisis." In *Ecology And Religion In History*, edited by David and Eileen Spring, 1203–7. New York: Harper & Row, 1974.

Whitehead, Alfred North. *Process and Reality: An Essay in Cosmology*. Edited by David Ray Griffin and Donald W. Sherburne. New York: Free, 1929.

Whitney, Barry L. "Divine Immutability in Process Philosophy and Contemporary Thomism." *Horizons* 7:1 (1980) 49–68.

Wiley, Tatha. *Original Sin: Origins, Developments, Contemporary Meanings*. Mahwah, NJ: Paulist, 2002.

Wilkinson, D. A. "The Revival of Natural Theology in Contemporary Cosmology." *Science and Christian Belief* 2 (1990) 95–115.

Williams, Daniel Day. *What Present-Day Theologians Are Thinking*. New York: Harper & Row, 1959.

Williams, Robert R. "Sin and Evil." In *Christian Theology: An Introduction to Its Traditions and Tasks*, edited by Peter C. Hodgson and Robert H. King. Philadelphia: Fortress, 1982.

Wogaman, J. Philip. "The Doctrine of God and Dilemmas of Power." In *Trinity, Community and Power: Mapping Trajectories in Wesleyan Theology*, edited by M. Douglas Meeks. Nashville: Kingswood, 2000.

Wolterstorff, Nicholas P. "Genius of Reformed Liturgy." *Reformed Worship: Resources for Planning and Leading Worship* 2. Online: http://www.reformedworship.org:magazine:article.cfm?article_id=24.

Wood, Laurence W. "From Barth's Trinitarian Christology to Moltmann's Trinitarian Pneumatology." *Asbury Theological Journal* 55:1 (2000) 51–67.

World Council of Churches. *Baptism, Eucharist and Ministry: Faith and Order Paper: 111*. Toronto: Anglican Book Centre, 1982.

Wright, J. Eugene, Jr. *Erikson, Identity and Religion*. New York: Seabury, 1982.

Wright, Wilmer Cave, translator. "Julian, letter 22." In *Julian Emperor of Rome: The Works of Emperor Julian*. New York: Putnam, 1923.

Yong, Amos. "The Turn to Pneumatology in Christian Theology of Religions: Conduit or Detour?" *Journal of Ecumenical Studies* 35:3–4 (1998) 437–54.

Young, Frances. "Creatio Ex Nihilo: A Context for the Emergence of the Christian Doctrine of Creation." *Scottish Journal of Theology* 44 (1991) 139–51.

Zikmund, Barbara Brown. "The Trinity and Women's Experience." *Christian Century* 104:12 (1984) 353–56.

Zizioulas, John D. *Being as Communion: Studies in Personhood and the Church*. Crestwood, NY: St. Vladimir's Seminary Press, 1985.

www.ingramcontent.com/pod-product-compliance
Lightning Source LLC
Chambersburg PA
CBHW051743230426
43670CB00012B/2146